Digital Prohibition

Digital Prohibition

Piracy and Authorship in New Media Art

CAROLYN GUERTIN

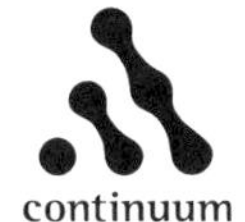

Continuum International Publishing Group

The Tower Building   80 Maiden Lane
11 York Road   New York
London   NY 10038
SE1 7NX

www.continuumbooks.com

Library of Congress Cataloging-in-Publication Data
A catalog record for this book is available from the Library of Congress

ISBN: HB: 978-1-4411-0610-0
PB: 978-1-4411-3190-4

Typeset by Fakenham Prepress Solutions, Fakenham, Norfolk NR21 8NN
Printed and bound in the United States of America

Contents

Acknowledgments

My journey toward completing this book was far from solitary. I had the company of students, colleagues, friends and family. I could not have arrived at this point without them. I would like to thank the Social Sciences and Humanities Research Council of Canada for two generous grants: a SSHRC doctoral fellowship (1998–2000) at the University of Alberta and a SSHRC Postdoctoral Fellowship (2004–6) at the McLuhan Program in Culture and Technology at the University of Toronto. The Provost's Office at the University of Texas at Arlington also generously recognized me with a coveted faculty development leave in the fall of 2009, and the College of Liberal Arts backed it up with travel funds. Those funds and acknowledgments of my work were accompanied by the very real support of Professors Pete Smith, Michael O'Driscoll, Michael Joyce, Kevin Gustafson and N. Katherine Hayles. I've also been blessed with friends in my chosen field as well, and I want to thank Maximiliaan van Woudenberg, Renée Turner, Rita Raley, Kate Pullinger, Sue Thomas, Talan Memmott, and Jésus Octavio Élizondo Martinez for their cheers, ears and critiques. I am indebted to artists who graciously shared their time and insight in interviews -- Steve Gibson, Khaled Hafez, Atteqa Malik, Mez, Huma Mulji, Feng Mengbo, Miao Xiaochun, Yifang Wang -- and to Melissa Chiu, who took time out of her busy schedule to meet with me. Thanks are owing too to the kind artists who made their images available for this book. I am grateful to my willing guinea pigs, my English 5380 students at UT Arlington and my Transart Institute students in Berlin, who let me test out my ideas on them first in graduate seminars, and especially to my graduate research assistants, Joy Sterrantino, Tricia Dupew, Jared Chambers and Johansen Quijano. I particularly want to thank my editor, Katie Galoff at Continuum, along with her team, for their patience and care, and for making the production of this book so fast and painless. My family, John, Irene, Nancy, Nick, Sarah and Ed and Bernice honor me by always being there. Finally, I am most indebted to Katherine Jin for her inspiration and support and for all of the many things she does to make my life easier every day.

Introduction
Ambivalence and authorship

Perhaps it was nowhere more apparent that capitalism and the youth of the Twentieth century are at odds than during the third week of September 2011 when the *New York Times* and other major American news media outlets failed for the ninth straight day to publish anything significant about the occupation of Wall Street and surrounding areas by a throng of more than 5000 strong. While the *Times'* front-page headlines trumpeted that "A Brutal Afghan Clan Bedevils the U.S." (September 25, 2011), it buried a small article on New York events with the headline, "80 Arrests as Protesters March in New York" on page 18 that tells of protesters being herded "like cattle" in the streets of Manhattan (Moynihan, 2011a, 18). Meanwhile, the New York Police Department's extreme use of force – including tackling participants and macing kettled protesters – on a vocal but peaceful gathering was treated as if it were an everyday affair. For young people, the perfect storm had arisen with the combined events of bank bail-outs, the ongoing mortgage crisis and the execution of an African-American man named Troy Davis, for a crime it is widely believed he did not commit. The role of the *New York Times* altered again on October 1, 2011, after 700 protesters were arrested for walking on Brooklyn Bridge, when the newspaper changed its online edition to shift blame off the NYPD and on to the protesters. By all early accounts, the police allowed the marchers on to the bridge and then herded them on to the roadway, where they were immediately arrested for an assortment of violations under city laws. The *Times* report at 6:12 p.m., under the headline "Protesters Arrested on Brooklyn Bridge," read: "After allowing them onto the bridge, police cut off and arrested dozens of Occupy Wall Street demonstrators." At 7:10 p.m., the headline became "Hundreds Arrested on Brooklyn Bridge" and the lead was changed to read "In a tense showdown over the East River, police arrested hundreds of Occupy Wall Street demonstraters after they marched onto the bridge's Brooklyn-bound roadway." The outrage from the change was long and loud on a wide selection of social media outlets. While the lack of coverage of the initial protest in the mainstream press is cause for deep consternation, blame juggling and slant represen-tation of events is a different matter entirely. Part of the reason for this is deliberate: the non-violent nature of the occupation does not lend itself very well to news coverage. While there are fears that web coverage threatens to tip toward only sensationalizing violence, it is clear that unless we see things with our own eyes we no longer know who to trust. The reality is complex and nuanced. At the event's Facebook page (with the hashtag #occupywallstreet), the mandate is described as being to "zero in on what our one demand will be, a demand that *awakens the imagination* and, if achieved, would propel us toward the radical democracy of the future" (emphasis added). The event grew out of the mandate of a group called "'General Assembly' (GA) – a leaderless, consensus-based decision-making process" (Schneider, 2011) and a three-week sit-in in a City Hall by the group New Yorkers Against Budget Cuts. #occupywallstreet was initiated by Adbusters, the Canadian anti-capitalist cultural organization, which floated a meme that called for an occupation of Liberty Plaza Park for an expected period of several months as a protest of the dysfunctionality of the current economic system. That, in turn, was taken up by Day of Ragers and Anonymous, and other militant groups. But what is most surprising is that it is not simply radicals or militants who have occupied the square, but the fact that they are largely absent. Instead the event seems to be manifesting itself primarily as an expression of justified outrage by ordinary people. Indicative of widespread disaffectation: "More than demanding any particular policy proposal, the occupation is reminding Wall Street what real democracy looks like: a discussion among people, not a contest of money" (Schneider, 2011). On October 1, they published the first edition of their own newspaper called *The Occupied Wall Street Journal,* with a print run of 50,000. Financed as a Kickstarter initiative, they raised $12,000 within the first eight hours (Moynihan, *New York Times Blog,* 2011). What is at the heart of this event, Schneider argues, is community formation and a creative outpouring of ideas from everything around how to protest to how to make posters to how to reinvent government. The crowd keeps chanting, "This is just practice." Time will tell what they are practicing for.

This is not just an American crisis, although OccupyLA, OccupyChicago, OccupyAustin, and OccupyDenver movements have spread since the Wall Street occupation began. Similar youth protests have occurred in squares in London, Spain, Greece, Tunisia, Libya, Syria, Bahrain, and in Tahrir Square in Egypt that have been organized and strategized through social media. As youth in country after country in the past few years have risen up to start their own Arabian Spring against repressive governments, or to protest how rotten capitalism is, or the fact that corporate interests wield more and more power and control more and more property, or that the youth of today are crippled by debt, youth seek to proclaim their right to author their own futures. One of the Wall Street protesters, Marissa Holmes, a documentary filmmaker and social activist, said, "We are the over-educated and under-employed. Our future has been totally sold out. Politicians have failed us and the [Union] square is somewhere where we can speak out. This is the beginning. It's direct democracy in action." (*Guardian*). These swirling, seething groups of humanity, according to Michael Hardt and Antonio Negri, are the multitude,[1] a movement or segment of society that does not simply just produce things in common, but produces *the common* – or what used to be called "the commons" in collectivist (read pre-capitalist) days (2004, xv). Diversity is the trademark of such a group, and it cannot be situated or identified on the basis of class, creed, color, or even politics. Through this escalation of technological expertise, social intellect and sophisticated peer-to-peer organizational structures, the "general intellect" (as Marx would have it) of these mobs becomes the crucial factor in production for this new phenomenon dubbed a "creative class" (a label misapplied from Tony Blair's Britain for a group that is neither a class nor 'producing' in the usual sense of the word). This group cannot be broken up since they are a networked and fluid body, and so the establishment tries to break them down into deviants and criminals to dismiss, discredit and disarm them: marginalized through queer politics such as ACT-UP and Queer Nation, anti-capitalist at demonstrations in Seattle and Genoa, and more recently blatantly ignored in the traditional press during the March on Wall Street, these cries for authorship are seen as "chaotic and incoherent" rather than part of a continuous narrative (2004, 192). Meanwhile, the multitude, assisted by technology, is getting better and better at organizing.[2]

What the multitude does on a daily basis is to churn out "constant innovation" (Hardt and Negri, 2004, 193) – equal to 48 hours of YouTube video every minute, for instance (in 2011, Rubel) – in order to produce the social sphere that we inhabit and enjoy. Capitalism is the Executive-Act vampire that keeps this Creative-Act group alive so that the former can continue to use and to feed off the latter. Originally, the intention in the privatization of intellectual property was to protect and encourage creativity and innovation. Instead, through counter-intuitive copyright 'reform' that privileged existing copyright holders over individual creators, corporate interests now have a stranglehold on private property and the public domain. The privatization of creative works and declining access to information means that not just creativity and innovation are being stifled, but a whole generation is being effectively silenced as it is told that it is no longer legal to use the raw materials of culture to create. They are not taking it lying down, as we have seen. According to Hardt and Negri, the surplus 'wealth' of intelligence, experience, knowledge and desire will be the catalyst for change. They say that "When communication is the basis of production, then privatization immediately hinders creativity and productivity" (2004, 185), since it is the multitude that is the very life force of a society. As private property has destroyed the notion of things shared and used in common, it is the force of the multitude's creative energies applied in "communication, collaboration and cooperation" that is "the primary characteristic of the new dominant forms of labour today" (2004, xv). "Anyone who works with information or knowledge … relies on the common knowledge passed down from others and in turn creates new common knowledge" (2004, xv). This is a fact that is brought home to us more every day as youth uprisings around the globe are rewriting governments (or trying to do so) from the ground up. But as digital technologies make it easier and easier to organize demonstrations, they are also facilitating a new model of creativity based on the ascendant model of production in the Twenty-first century: the remix.

The third space of authorship: participatory practices and new narrative models

In the influential book *Location of Culture* (1994), Homi Bhabha uses the concepts of mimicry, interstice, hybridity, and liminality to argue that cultural production is always most productive in spaces where it is most ambivalent. A major site of ambivalence in the realm of digital art and literature lies in the fact that so much of this work exists outside of the capitalist economy of exchange and commodified culture. Authorship itself, by definition, is a declaration of property and an economic claim. Digital work, on the other hand, occupies the space of what is defined as "free culture": not a space without property, but a space of the free play of ideas. (In fact, in reference to Homi Bhabha we might think of this as the third space of authorship: a site of resistance and renegotiation.) In cyberspace everyone is potentially an author, and, in fact, in a user-generated culture, arts exist not at the site of the iconic author-genius laboring in isolation, but at a social interstice that might provide a collective model – a model that evades or attempts to evade the pitfalls of consumer culture, commodified objects, and monetary exchange. I will explore new creative practices, media as a vehicle for social reform and aesthetic expression, and some of the potentialities in the social nature of electronic works – literary, artistic, and viral – to create new kinds of creative practices, and new spaces for the rise of alternative artistic, authorial, or publishing models.

Authorship is a collaborative act. In fact, prior to the Nineteenth century in the West all authorship was seen to be a collective exercise, with writers standing on the shoulders of the giants that came before them. Even in our contemporary world, editors, publishers, critics, mentors, literary muses, and collaborators all form a part of the web that is the text, revealing the lie in the myth that the book is a solitary act. Andrew Bennett argues that it was Romantic ideology – most conspicuously put forward by William Wordsworth – that embedded the author as an owner of the work, a concept that would ultimately become enforceable in law (Bennett, 2005, 100). Compare the literary author model to a film production crew or to the creators of a medieval cathedral or a video jockey at a rave, and we can see how the whole idea of true collaboration becomes impossible within the confines of the literary volume as we have known it. In the art world too, gallery owners tend not to be interested in collaborative artworks because it is the single name on the *objet d'art* that sells.

It is in the postmodern world that this Romantic construct becomes troubled. Roland Barthes proclaims the 'Death of the Author' when he redefines the text as a "multi-dimensional space in which a variety of writings, none of them original, blend and clash" (Barthes, 2005, 15). For Barthes, the text is a negative space and the text-as-intertext (that is to say, a text interrelated to other works) eliminates the central, controlling power of the individual from its very conception. Michel Foucault in his answer to Barthes' piece "What is an Author?", sees the author function also in spatial terms, as "a space into which the writing subject constantly disappears" (Bennett, 2005, 20). According to Foucault, this author function emerges as a principle of ownership – or possessive individualism – in the Eighteenth century with the arrival of copyright law (Bennett, 2005, 23). Individuality's arrival can also be tied to the technology of the printed book, which helps birth public and private space just as surely as the fencing of the English Commons produced enclosures.

The notion of the solitary author model makes sense when the book or work of art is seen as part and parcel of property values. Mark Rose in *Authors and Owners* argues that the modern concept of authorship:

is "inseparable from the commodification of literature", and that the "distinguishing characteristic" of the modern author is "proprietorship": the "invention" of British copyright in 1710 affirms the new conception of the author as the "originator and therefore the owner of a special kind of commodity, the work" (Rose, 1–2; quoted in Bennett, 2005, 50).

It is important to remember that copyright was *not* invented to protect authors; it was an innovation designed to guard the corporate interests of *publishers and printers* (Bennett, 2005, 51).

Since the Eighteenth century authors and artists in their turn have sought to transgress the boundaries of property and distance themselves from such earthly concerns as money. The starving poet in the garret is the ideal literary model in what Pierre Bourdieu dubs a reversed economic world. Literary aspirations elevate one above the worldly concerns of monetary needs: an author's "'cultural capital' is based on a *disavowal* of capital, of capitalism, of the economic or mercenary motivation for writing" and it is that disavowal that lends value to authorship (Bennett, 2005, 53). But where this notion of the solitary genius in print was a fiction that was not sustainable even within print culture, it is even more absurd in the networked spaces of new media. New media works are not property based, but fluid, deterritorialized systems in motion. What we therefore need – and what are evolving – are new literary and artistic models to understand the burgeoning nature of authorship in electronic works. As with so many other aspects of user-generated culture, these are works that are *audience-* rather than *author-*defined. For example, Douglas Rushkoff's collaboratively written experimental novel called *Exit Strategy* will not stand up to Web 2.0 scrutiny, regardless of its literary merits. Billed as the first 'open source novel', Rushkoff's story remained online for over a year before it was fixed in print with the 100 'best' footnotes and annotations from the Web included in the text. Open source is a design, development, and distribution philosophy that originated with software that calls for transparency in the bones of that which is being created. It invites users in to be a part of the project, working to customize or to improve the software for the community at large. The fact that Ruskoff's book invited the audience in may or may not have been as revolutionary as he claimed, since collaboration is key to any literary work; the fact that the novel was published as a commodity under Rushkoff's name, with the contributors' addenda remaining only as footnotes, rather than interventions in the text, tells you how close it stays to a traditional, print-bound novel.

California-based artist Nanette Wylde's electronic flipbooks are more truly interactive. Originating in the Nineteenth century, flipbooks were a primitive popular art form inspired by early experiments with film and the moving image. You may have seen them in low-tech versions which produce a moving image by flipping the pages of a book, and as such we can see them as a place where film, the comic strip, and storytelling meet. Thomas Edison's

Kinetoscope (1893) was one automated form and the Mutoscope (1895) was popular in peep-shows. The Kinetoscope was really quite primitive, only an automated step up from flipping pages by hand. Now Nanette Wylde has taken the idea of the flipbook and renders it textually as an interactive experience. Composed of databases of combined content *input* by users and by Wylde, *HaikU* and *Belief Manifesto* play upon the Nineteenth-century idea of images in motion to create revolving interactive texts instantly. Drawing from three different datastreams, these texts recombine them to create instant works of art from a community experience. *Arrested Movement* works along similar lines but more closely shows a genealogy with text-based experiments by feminist artists like Jenny Holzer and Barbara Kruger. Exploring our expectations of social posturing and the politics of rhetoric, *Arrested Movement* toys with our expectations. Combining racial and ethnic groups with the politics of everyday activities, the texts consistently test our assumptions and challenge us to rethink them. While *Arrested Movement* does not require the reader to input material, it does work to resituate the reader in relation to the material she is reading. In short, it requires us to change our perspective and insert ourselves into the textual community to experience new points of view.

The new user-generated culture that is emerging from the Web 2.0 revolution is creating new modes and forms for storytelling and all kinds of interactor engagement. In fact, one of the biggest shifts is that the new works are looking less and less story-like or art-like. Story, for example, is something that can now be harvested or generated. Stories can be tiny, encompassing something as small as an animated gif image that lasts only a fraction of a second. A number of automated storytelling engines that use Twitter social software and its community to tell stories retain both the spirit and the letter of Wylde's multi-streamed interactivity and combine them with user-generated practices. In fact, Jenny Holzer herself now keeps a Twitter account, demonstrating how far ahead she is of the rest of us or, alternatively, to alert us to the fact that it has taken the world several decades to catch on to Holzering as a way of being in the world. Twitter is a microblogging software that allows instant messaging to private or public networks on any topic as long as the posts do not exceed 140 characters at a time. These microposts are known as "tweets" and their content from the general public generally runs from the banal to the mundane. In 2007, Twitter was voted the "Best of the Web" at the South By Southwest Interactive Festival in Austin. That same year, the *New York Times* acknowledged that Twitter was "one of the fastest-growing phenomena on the Internet" (Pontin, 2007). By 2010, according to *Wikipedia*, more than 65 million tweets per day were being sent. Since its inception Twitter has gained fame for its resourcefulness as a civil rights tool; it is also known for its versatility in natural disasters (e.g., the Haitian and Chinese earthquakes in 2010 and 2008 respectively) and dangerous situations (e.g., the Virginia Tech massacre). Twitter is also being put to work as a new engine for storytelling. *Twistori* is the most innovative Twitter-derived fiction experiment to date. It uses a Java applet co-designed by Amy Hoy of Maryland and Thomas Fuchs of Austria to scoop content from Twitter. The posts it harvests all contain one of six different phrases: "I love …," "I hate …," "I think …," "I believe …," "I feel …," and "I wish …" Using Summarize (a Twitter-based conversational search engine) and WeFeelFine (a Java applet that harvests emotional content from blogs), it creates serial poems, or serial stories, based around basic in-the-moment human expressions of emotion. It is compelling to watch. As Sol LeWitt said, "The idea is the machine that makes the art" (quoted in Peretti, 2007, 161). The authors have created the machine that harvests the content, but create no new literary content themselves; the aesthetic attributes of the work are in the *process* of data collection and playback rather than in the *creation* of new content.

Are these experiments literary? Not in the least, although a certain poetry arises in the 140-character brevity as with the Haiku form. Bruce Sterling, the science fiction writer and journalist, is reported to have tweeted, "Using Twitter for literate communication is about as likely as firing up a CB radio and hearing some guy recite 'The Iliad'" (Pontin, 2007). However, just because it is not literature does not mean that it is without literary or aesthetic value. What are being enabled here are new interactive models of spontaneous storytelling that are more akin to orality. 'Literature' (the word and the concept) is fixed in the foundations of print; storytelling on the other hand is a more organic, oral form of narrative. Just as the notion of solo authorship is an uncomfortable inhabitant of the property-less, deterritorialized digital space, so too the literary writer of days of yore is a bit out of his or her milieu here. These new anonymous arrivals are occasional writers who write to feel connected, to rise to a challenge to make themselves heard within a community of like-minded social beings, who tweet because they can. Out of such simple social interactions complex revelations – the secret guilts, intimate thoughts, philosophical ponderings – of a living, breathing surge of humanity is made visible before our eyes. As a manifestation of sociability and constructed relations, it may be a new model for a living art form. It is certainly a model of distributed authorship and I would argue that this is a manifestation of authorship as popular culture in motion.

For Nicolas Bourriaud, these new electronic literary forms would be an example of relational aesthetics. Relational aesthetics is: "a bona fide political project because it makes *social relations*, rather than art a political issue. These … relations seek to reconnect people and work against the atomization of the individual in a mass media society" (Pennings, 2005). No art object is present here; instead, what gets privileged is Bourriaud's cluster of relationships: proposals, dialogues, encounters, and negotiations. As conceptual artists like Jenny Holzer have tried to demonstrate all along, these new rhetorical positionings for art and art objects may be integral to class, gender, race, and postcolonial politics too. The object (what we have read) has not entirely disappeared, but it is not the focus or the reason for this new generation of stories either. Even if we could archive these online stories (and the Library of Congress is), would we want to? Is their meaning not in their presence and active sense of community creation *in real time*? The creative work becomes a space of encounter for Bourriaud, where contact and tactility are privileged over the visual: "The space where [these] works are displayed is altogether the space of interaction, the space of openness that ushers in all dialogues" (Bourriaud, 2002b).

In this same way, the interactive installation *TypeTrace* seeks to engender a communal writing and socially networked experience that mirrors the users' thinking about writing. The brain child of Paris-based Japanese composer-turned-media-artist Takumi Endo and engineer Shinya Masuyama, *TypeTrace* questions what it is to be an author, a reader, and a member of the audience. Endo explains:

> Every day, we read dozens of emails, some newspapers and magazine articles, several advertisement billboards and maybe a few parking tickets. We are surrounded by letters perpetually. But all these printed words and fonts are literally "dead still," just like rotten corpses. There is no wonder we feel so relieved when we receive hand-written notes from colleagues, or letters from our lovers; we are suddenly reminded of the richness of the pen strokes; they give you such abundant resources for guessing how, and under which mood and feeling the message was produced. The strokes are the faces of people we have forgotten a long time ago after being flooded by digital messaging tools; they let us imaginarily revive the generation of the text we face (Endo, 2007).

Reawakening the author function within us by giving us back the tools to connect, the mechanization of a keyboard at the front of the room echoes the sound of creation as interactors sit at laptops and make their thoughts visible on the screen before them – with added emphasis. If one hesitates, words take up more space. If one corrects a typo, the ghostly mistake is preserved along with its replacement. The human hand is thereby reinserted back into the authorial process. Not only is the personal physically present in the installation, but the Kinetic Keyboard software has been released as open source code under a Creative Commons license for programmers to adapt to their own ends. In this way, text is thereby being reconnected to context and the practicalities of production; the message remains personal and variable while the system becomes an organic and communal experience.

Similarly, Marie Sester's playful digital installation, *Access*, which debuted at Ars Electronica in 2003, is about community formation more than it is about story as we are used to exploring it. The spotlight, that seeks out a person in real space, is triggered by real Web users in real time; the audio track is sampled from pre-recorded snippets and is often concerned with issues of surveillance. People who are targeted try to outrun the spotlight, try to transfer it to other people or perform for it. The playful or fearful stories that arise from it are personal narratives that engage us, like play. Perhaps not surprisingly the ideal model for interactive narrative has been defined by Janet Murray as precisely this activity: children's imaginative games.

Instead of following a top-down model, these forms of engagement affect us so deeply because they function as peer-to-peer networks. "Peer-to-peer systems, like other distributed systems, are like Punk rock: They empower fans and citizens, create new communities, and close the gap between creators and consumers" (Vaidhyanathan, 2004, xvi). We might think of this, to adapt Homi Bhabha's terminology, as the *third space* of authorship. The first two constructs are polar opposites and, as such, are entirely at odds. The originary or first space of authorship is the oral storyteller. She transforms lessons, history, wisdom, and traditional tales into compelling communal events. The second space of authorship (or artistry) would be the now-endangered Romantic idealized solitary genius. The third space I would propose as a fluid site for the construction and reconstruction of authorship, a collaborative zone where it is possible to speak of resistance, appropriation, negotiation, contingency, translation, otherness, commonalities – you name it. Trying on and discarding masks in this new space allows ambiguity, complexity, and hybridity to play out within new aesthetic practices (which I will explore in the next chapter). This new space can open up possibilities for the invention of new social structures of narrative authority and for new interpretations of the contingencies of temporary and fluid authorial identity – and new tactical behaviors – that the earlier models of authorship on their own never made possible before.

Copyright evolved independently from authorship where it was first designed to protect a physical object (the book), then extended to protect 'the work' and now, as restrictive corporate powers increasingly have their way, is being used to restrict access to digital information: the very building blocks for new creative works (Sundaram, 2007, 60–1). James Boyle likens the current trend to a second enclosures movement as changes in copyright and patent law are restricting access to what he calls "a commons of the mind" (Boyle, 2005, 237). In the first enclosures movement in Britain, the fencing of common land led to far more efficient methods in farming, to industrialization, and to a massive workforce migration from rural to urban spaces. This was based on the reality that land was scarce and could only be used for one purpose at one time. Current changes to copyright law, however, falsely posit that intellectual property is a similarly scarce resource. Boyle points out that, unlike a plot of land, a "digital text … can be used by countless people simultaneously without mutual interference or destruction of the shared resource" (2005, 241). Witness Nanette Wylde's work or Ian Hatcher's 'open source' texts, which allow for play, experimentation, and communal authorship via its multiple users. Or think of viral media which spawn multiple new versions – like the Hitler rant meme with more than a thousand variants on a single scene of Oliver Hirschbiegel's *Downfall* (*der Untergang*; 2004) uploaded to YouTube. And a Web 2.0 text does not have to be high-tech for the user either. Nanette Wylde's *Meaning Maker* consists of personalizable downloadable documents. There are six different forms, including *American Citizenship, Art Viewing Experience*, and *The Family Gathering*. She advocates that you use these forms to enhance your experience of live events and to lend a tongue-in-cheek evaluative structure to your undertakings. *The Academic Conference Edition* allows you to rate speakers, your own performance, your dress, your inadequacies, your state of mind, and the quality of your experience. The point of such a conceptual art project is that since it is infinitely renewable, it functions as a social critique of the expected norms of particular kinds of experiences. But as conceptual art, it also does not fit the finite, quantifiable 'property' model that copyright assumes. As Tarleton Gillespie puts it, a cultural object such as a "book maybe exhaustible, but the 'work' itself is not. Culture cannot be consumed" (Gillespie, 2007, 25).

Instead of plentitude or efficiency, what these sweeping changes to copyright are engendering are a shrinking territory of materials for creative minds to draw from as sources of inspiration – and an increasingly militant body of creative practitioners. In China, a perfect counterfeit copy of an UNESCO heritage Austrian town, the town of Hallstatt and its lake, is being re-created by a corporation in the province of Guangdong. The legalities of the re-creation are being explored by Austrian officials. This is not the first incidence of such an event. The city of Anting, 30 kilometers from Shanghai, re-created a medium-sized German city, including "Bauhaus style architecture and a fountain with statues of Goethe and Schiller," and:

> In 2005 Chengdu British Town was modeled on the English town of Dorchester. One year later Thames Town was finished near Shanghai,

complete with a 66-meter tall church that bears a striking resemblance to a cathedral in Bristol. Also near Shanghai are mini versions of Barcelona, Venice and the Scandinavian-inspired Nordic Town. The architectural plagiarisms are popular destinations among middle-class Chinese, even serving as backdrops for wedding photos (*Spiegel Online*).

In Berlin, members of the Pirate Party were elected to office in September 2011, after an "irreverent campaign captured the imagination of young voters as the party expanded its platform from an original focus on filesharing, censorship and data protection, to include social issues and citizens' rights" (Dowling, 2011). In the past, the argument for creating an intellectual property right law has been that "without an ability to protect their creations against theft, creators will be unable to earn an adequate living. There will be inadequate incentives to create" – or so it has been claimed (Boyle, 2005, 241). Of course, all creative work builds on the past. The Mickey Mouse Act (which extends copyright every time Disney's creature is close to expiring) is built on a denial of the fact that the Walt Disney empire was founded on the appropriation of properties like the Brothers Grimm fairytales and Buster Keaton's Steamboat Bill character as source material. The underlying assumption is not only a privileging of the established author over the new, but also the myth that *only a select few in society are practicing creative acts.*

Web 2.0 sites like YouTube, and the open source and free culture movements demonstrate the lie behind this rationale. In the early days of computing, software was free. Like the realm of ideas, anyone was able to tinker with and alter or improve the code for their own purposes. Gradually in the 1980s this started to change as proprietary software became the norm. In response to this threat, MIT researcher Richard Stallman started to work on the creation of a free operating system, which he called GNU. GNU later joined with Linus Torvald's Linux to become the leading open source operating system. What is revolutionary in Stallman's system is the concept of licensing, what he called a General Public License (GPL) to ensure that software remained free:

> Software licensed under the … GPL cannot be modified and distributed unless the source code for that software is made available as well. Thus, anyone building upon GPL's software would have to make their buildings free as well. This would assure, Stallman believed, that an ecology of code would develop that remained free for others to build upon. [Stallman's] fundamental goal was freedom; innovative creative code was a byproduct (Lessig, 2004, 280).

The creative act by every user thereby becomes an essential act: as essential as breathing. Within the Free Software movement everyone is an author.

Lawrence Lessig's Creative Commons and Free Culture movement is built on the same premise:

> an argument for free culture stumbles on a confusion that is hard to avoid, and even harder to understand. A free culture is not a culture without property; it is not a culture in which artists don't get paid. A culture without property, or in which creators can't get paid, is anarchy, not freedom. […] A free culture, like a free market, is filled with property. It is filled with rules of property and contracts that get enforced by the state (Lessig, 2004, xvi).

So where does this leave the deterritorialized writer of digital narrative or author of creative works, laboring in the as-yet-still-free space of the Web? New mobile interfaces like the iPad are built with copyright restrictions and Digital Restrictions Management (DRM) in mind, and, as a result make it much harder to be creators. Many of the tools that are available for harvesting material from the Web and other sources are not functional or have no equivalent alternatives (yet) on the iOS. With Apple's stranglehold on what gets approved and the grandfathering of digital locks on previously purchased content, it seems unlikely that remixing or sampling will get any easier within this operating system. The practice, however, continues unabated, as remixing, sampling, and reusing found material become the primary modes of creation. Apple's announcement of a coming Android unit also does not bode well for creative types who want to work with an OS for existing materials given the locked-down nature of the iOS, an operating system where every app must be approved by Apple to be released. The digital artwork is a space of encounter, where community, contact, and tactility are privileged over the material object: "The space where [these] works are displayed is altogether the space of interaction, the space of openness that ushers in all dialogues," according to Nicolas Bourriaud (2002). He also argues that artists are able to provide non-commodified experiences because they operate in a "social interstice" that evades capitalistic impulses. "Art represents a barter activity that cannot be regulated by any currency, or any 'common substance'." It is the social nature of relational aesthetics that frees it from the tyranny of the object/commodity (Bourriaud, 2002, 161). Much of the music business is built on this free model: give away your songs on disc, so that you can live off the revenues of your performance career. This is not to say that we ignore such earthly concerns as paying rent or buying groceries, but – since there is no scarcity of digital resources and their sum cost is virtually free – we need to find alternative ways to help digital art support artists. Digital expertise can place new media artists in other professions that free them/us to license work or generate grants that are not strictly speaking part and parcel of the capitalist system, but that allow authors free play: the ability to create while making a living. This is not the best system, but in this dark age of copyright restrictions it is a necessary one to help keep the creative process functioning as a social interstice that can engender new models. In order to keep the freedom of creation alive, it is important that we, as Twenty-first-century conceptual artists and authors, develop new collaborative models – new play spaces. In fact, conceptual models and activists' tactics already among us may just be the new practices and tools we have been looking for as alternatives to the slippery slope of a consumerist culture that thwarts creative practice and stifles fresh ideas.

The music industry has been steadily dwindling for a decade since the advent of Napster in 1999, but the indie music scene is alive and well. Sites like ccMixter now pose new possibilities for publishing and distributing creative work. While the book trade's future is a dark page at this point with so many publishers in North America going under or laying off staff, and the large bookstore chain Borders going broke in 2011, they are not the first wave in the wake of change from newspapers to video rental stores. The design industry and font designers were the first to be felled by digital technologies. That sea change is now being felt across the board in every area of production, including manufacturing and prototyping as 3D printers and custom fabbing come to the fore. New kinds of creative projects, mobile media, locative media, and new modes of personal distribution and (self-)publishing are the future. The rise of locative media, the book app, and augmented reality hold interesting possibilities as new interfaces for digital art and literature. More significantly, as more and more is put within our reach via technology, the more it enables collaborative practices.

Look at the collaborative work, for instance, that is being created by Perry Bard in her public undertaking *Man with a Movie Camera: The Global Remake*. Perry Bard is a digital media artist based in New York City. Her site-specific public video performances strive to build communities around cultural, historical, and mnemonic issues, frequently by involving the audience in the production process of the work itself. This undertaking is a mammoth one, financed by the BBC, the Arts Council of England, and the Canada Council, and one that would not have been possible if Dziga Vertov's original film, which was made in 1929, was still under copyright. This work is something she calls "Vertov in the age of YouTube." An experiment in database cinema, this participatory project is infinitely expandable. The content itself is *not* made by Bard. What she has created is the software (which structures and assembles all film clips), the indexing of the original film, and an online portal to upload it.

Based on ... Vertov's 1929 masterpiece, this version takes his concept of 'decoding life as it is' into the twenty-first century. Blogs, YouTube and sites such as Facebook and Flickr have revolutionized the way we communicate and created an entirely new system of social networking (biggerpicture).

The pinnacle of Russian Constructivist film that glorified industry, the worker, and the commonalities of the everyday, Vertov's *Man with a Movie Camera* is considered by many to be the greatest film ever made. Much like web culture today, Vertov's approach was born of experimentation and his rejection of "the values of 'high' conceptual art. [To him,] [t]he artist was not an 'inspired visionary' (Bordwell 128), but rather an engineer using tools to create art" (Carriero-Granados, 2009). Vertov's footage was taken from filming in three different cities: Riga, Kiev, and Moscow. Resembling a documentary, the original silent film is structured more like a symphony than a narrative. It begins with an empty movie theater. Then, the projectionist sets up the reel, cut to the chairs, cut to the conductor and the orchestra held in suspension. The audience files in and then we follow the camera and the cameraman as they record the events of a day in the life of a city.

Perry Bard sets out to re-create this masterpiece not merely as an echo, but as a 'translation' into modern-day concerns that becomes a statement on contemporary culture and the Twenty-first-century urban environment. Part of that statement is that no single perspective will suffice any more, and so not only do we get the binocular vision of the old version juxtaposed with the new, but new shots are uploaded daily and rotated if there are multiple versions of single clips. This means that the film as it unfolds may well never play the same way twice. Vertov's

> film begins with titles that declare it "an experiment in the cinematic communication of visible events without the aid of intertitles, without the aid of a scenario, without the aid of theater." It is often described as an urban documentary yet the subject of the film is also the film itself – from the role of the cameraman to that of the editor to its projection in a theatre and the response of the audience. It is a film within a film made with a range of inventive effects – dissolves, split screen, slow motion, freeze frame – all of which are now embedded in digital editing software (Bard, 2011).

By making us the filmmaker, Bard extends the lens still further so that we as audience are encapsulated within its frame. We become the subject, just as Vertov's cameraman was the subject, by participating in the project as filmmakers. This is made even more explicit as Bard often shows the film as public art at public festivals and in spaces where the piece itself organically merges with the urban landscape that it set out to document.[1] The work is literally and figuratively infinite. It will continue to grow and change as long as people are sufficiently interested in it to continue to contribute to it. Vertov's masterpiece – which seems so contemporary to us now that it might not be out of place as an MTV-style video – was not well understood in his lifetime: "unable to get his films into theatres, he showed them in public spaces such as worker's recreation halls" (Bard, 2011). As such, Bard's use of the films as public art shows that they have come full circle. The question of who 'owns' the work does not even arise, although Bard is given the credit for the inspiration.

Free culture and public art are both about dollars and about liberties. Where copyright is about property, free culture is about freedoms – and this is not to be confused with open source practices. The split between free software and open source is on the issue of ethics. Free software was created to oppose the whole idea of Digital Restrictions Management (also known as Digital Rights Management – DRM) "as it is the antithesis" of free software (Stallman). Since the aim of DRM is to "trample your freedom, DRM developers try to make it hard, impossible or even illegal for you to change" that software (Stallman). Malicious features may include spying on users, restricting users, installing back doors, and imposed upgrades. Pressure from movie and music companies to use these tactics is particularly strong. (Think of the ironies of Elayne Zalis who stored her manuscripts and works-in-progress on *Scribd*, until the service removed them for copyright infringement. Who had she infringed? Herself, of course.)[2] Ever bought a CD or DVD that only works on some systems? That's DRM. Ever paid for a song on iTunes and tried to play it on something other than your iPod or your computer? You're in violation of DRM. Ever made a digital mixed tape? Both the DRM and copyright police will get you.

In April 2008, copyright laws changed in the U.S.A., quietly and without fanfare. Border guards are endowed with the powers of copyright police and have the right to seize any digital devices or files without suspicion of wrong-doing. For artists and academics this is a particularly dire situation because it even overrides Fair Use doctrines. Fair Use is not a privilege; it is a right. At what point can the border guard, in his newly defined role of copyright cop, distinguish between a legitimate copy for aesthetic expression and/or consumer use and a pirated one for commercial gain? These are the places at which the difference between Fair Use and Digital Restrictions Management become murky indeed and there are no easy answers. Changing the strangle-hold that copyright law has on creative works so that works are freed up for experimentation and creative play makes a lot more sense. By licensing creative works for alternative uses, we can open up creative gates that would have otherwise been closed by corporate restrictions. This might resonate for us most clearly if we think of it as a Twenty-first-century Prohibition.

The new Prohibition: digital piracy and the politics of creation

Dubbed "the noble experiment," the prohibition of alcohol was a grand attempt at social engineering in effect from 1919 until 1933 in the United States. However, while it began with the best of intentions arising out of a drive by the Women's Christian Temperance Union, the reality was that it was an abysmal failure. Born out of the desire to "reduce crime and corruption, solve social problems, reduce the tax burden created by prisons and poorhouses, and improve health and hygiene in America" (Thornton, 1991, n.p.), the new laws did initially reduce alcohol consumption. That period was brief though, as forces gathered to turn the ban into an opportunity. Prohibition fostered the birth of organized crime, greatly increased the dangers of the consumption of alcohol as safe standards of production disappeared, stretched the court and prison system to "the breaking point," bred contempt for laws and law enforcement, and brought about widespread corruption among public officials (Thornton, 1991, n.p.). Bootlegging, which was the illegal sale of alcohol as a beverage, flourished, as did the number of illegal watering holes known as "speakeasies" or "blind pigs." My great-grandfather was a bootlegger.[1] He ran the concession for the billiard hall in the basement of the famous Windsor Hotel in Montreal, which in those days was an international playground of the rich and famous. Canadian prohibition played out a little differently than in the U.S. In general it was not the making of alcohol that was prohibited in Canada but its sale in public establishments;

possession and consumption continued to be allowed in homes in private. Unlike in the U.S.A., alcohol continued to be available for purchase "through government dispensaries for industrial, scientific, mechanical, artistic, sacramental and medicinal uses. Distillers and brewers and others properly licensed could sell outside the province" (*Canadian Encyclopedia*). In Canada, the easiest way to procure liquor was through family doctors who would write prescriptions. The rash of epidemics was apparent in the fantastically long line-ups at doctors' offices at Christmas time. When my great-grandfather's illicit activities were discovered, officials at the Windsor Hotel showed him the door. For many in those days, unemployment meant destitution, but to cease being a bootlegger meant that he was freed up for the far more lucrative business of being a rumrunner. Rumrunning was smuggling across the border into the United States. This became easier and easier as Prohibition only lasted a year in Quebec, and the American market was just a few miles away and very, very thirsty. More than that, where Prohibition had been designed to protect people from alcoholism-related diseases like cirrhosis of the liver (a condition that is fairly rare among alcoholics), ended up instead by causing 50,000 deaths and uncountable cases of blindness and paralysis from bad liquor. The liquor tax in the First World War had generated many times more revenue for the U.S. government than War Bonds ever did. Moreover, the administrative costs of Prohibition were astronomical. The U.S. government allocated $5 million dollars for enforcement in 1919; by 1933, real spending had escalated to $300 million. Anyone could set up a still in their home. An estimated 30,000 Cleveland residents sold liquor during Prohibition, but it is believed that an additional 100,000 made moonshine or bathtub gin for themselves (*Digital History*). The consensus was that "the 18th Amendment resulted in 'evil consequences' … and it was said 'that Prohibition created an orgy of lawlessness and official corruption'" (*Digital History*).

History has a way of repeating itself. As we now sit perched on the verge of what may be another great depression, there is a new Iron Law of Prohibition that has been passed. It is now legal in the United States for border guards to search and seize your files, your laptop or iPod, or any other device or property to check for possession of pirated material. Similar laws are on the table in the U.K. and EU, but have been greeted with hostility in Canada. This is not an isolated phenomenon. In 2001, Russian programmer Dmitry Skylarov was arrested, imprisoned, and then held on bail in California under the Digital Millennium Copyright Act (DMCA). Matthew Fuller notes in his article "Commonality, Pixel Property and Seduction: As If" that Skylarov's crime was co-writing a hack that allowed any user to access the password to the encrypted Adobe 'portable document format', a PDF. It is illegal, under the DMCA, to "distribute 'circumvention technology'" – even if you just want a way to access your own forgotten password (Fuller, 2011). "Within this wider context," he continues, "what are the protocols that artists work on? Firstly, we must say that the conflict between commonality and property is never finished" (Fuller, 2011). Criminalization is increasingly the path of choice for copyright enforcers, as Jammie Thomas found out to her horror. A native-American mother of four from Minnesota, she was charged with violating copyright after illegally downloading 24 songs. (More about the legalities of downloading shortly.) Because she refused to settle out of court and pay the absurd fines, hers was the first case tried before a jury on the basis of new, much stricter copyright laws in 2007. She lost and was ordered to pay more than $200,000 for infringement. A new trial was called, due to errors in the first case, and in 2009 Thomas lost again. This time she was ordered to pay more than $1.9 million in damages to the record labels – a sum reduced by a judge to $54,000. The record labels appealed the reduced award and went to trial again, and were awarded $1.5 million by a jury. The court again reduced the fine in 2011, this time to $2,250 per song. The record companies are again appealing. Should someone lose everything for illegally obtaining two dozen songs? What would the penalty be if she had stolen two or three CDs from a store? How can this possibly make sense? Whose interests are served by these acts being criminalized? The escalation continues as creative tools, research methods, and technologies are outlawed. The notorious case of the jailing of researcher and hacker Aaron Schwartz was also international news. A man with a history of creating free software, criticizing corrupt corporations, and laboring on behalf of nonprofits, Schwartz is the creator of watchdog.net, a site that facilitated people accessing their own government-held data, and of Open Library, which made information on millions of books available to the public. Having been caught and arrested for data theft after downloading four million files from the Massachusetts Institute of Technology and the academic database JStor to use as a large dataset for research, he was imprisoned. The subject of his research is open to speculation, but he has a history of studying how money corrupts corporations. That he broke the law is not disputed. Yet the severity of his possible sentence is off the map; he could spent up to 35 years in prison and face $1 million in fines (Humphries, 2011). Why should we care? Bootlegging whether it is liquor or music is illegal. People who break the law deserve what they get, surely? But when we take a look at what exactly is being defined as piracy in our time, this issue is extremely complex and is increasingly being redefined in ways that are beneficial to the profit margins of corporations. Digital data are *designed* for downloading and sharing, and computers facilitate *the easy copying and altering* of digital files. That is the technology's goal (combined with the storage and searchability of those materials).

Since its inception in the United States since 1790, fines and penalties have not only escalated but have, as we have seen, been criminalized. According to the *Harvard Law Review* in "The Criminalization of Copyright Infringement in the Digital Era," copyright was first criminalized in 1897. At that point the concept of intent – "willful and for profit" – was introduced into the copyright infringement legislation. No fines were mentioned, but criminal prosecution with jail time of up to one year was introduced. In 1909, fines were introduced of between $100 and $1,000 and jail terms of up to one year for an expanded definition of copyrighted materials with the exception of sound. Sound was included in an extension in 1971. In 1974, penalties were raised for sound, as sonic recordings were deemed so lucrative as to require additional deterrence. In 1976, the entire Copyright Act was revised. This time much more severe penalties included an increased fine for infringement of up to $10,000; motion picture infringement of up to $25,000 and repeat offenders subject to fines of up to $50,000 plus two years in jail. Intention to make a profit was struck down and replaced by the "desire of financial gain" as a motive was introduced (1708). The 1976 Act also required that any existing copies and copying equipment be destroyed. In 1982, the Piracy and Counterfeiting Amendment Act was introduced. This increased criminal sanctions of copyright infringement for motion picture and recording industry violations. It also enhanced penalties for reproduction/distribution of sound recordings, motion pictures, and other audiovisual works. Large-run duplication (more than 100 copies during a 180-day period, and more than 65 copies of films in a 180-day period) is now classified as a felony, even for a first offense. In June 1992, software was added to the Act. In 1992, the Copyright Felony Act also came into effect. All copyright infringement became a felony once the threshold number of copies was reached – but the threshold was lowered to a mere *ten copies* with a value of more than $2,500. Jail time became five years and fines rose to a maximum of $250,000. Home computing software use increased exponentially in the 1980s and 1990s, and software piracy kept pace. In 1990, the loss in the software industry in the U.S. due to piracy was considered to be $2.4 billion. "Software was considered to be 'an easy target for thieves' because of its intangible nature, its relatively high retail price, and 'the fact that perfect copies of the most sophisticated programs can be made in seconds'" (1711). As a result of this, in 1992, software was deemed to require especially high sanctions. The great fear for the legal establishment was that most of this piracy was untraceable and unenforceable. The *Harvard Law Review* (1712) states:

The digital era has witnessed the rise of a new type of pirate, against whom it is often impossible to enforce criminal copyright provisions.

Anyone can now commit major copyright infringement because of the widespread accessibility of copying technology and the technology's ability to make perfect reproductions. Furthermore, the development of the hacker culture, in which people seem to abuse intellectual property as a means of challenging authority and showing off their expertise, has created a group of infringers who are not seeking to make a profit, thus making traditional criminal copyright laws inapplicable.

In April 1994, an MIT student, David LaMacchia, was indicted on wire fraud for hosting a bulletin board that encouraged members to share free pirate software – and that they had to the tune of about a million dollars in commercial ware. Because he clearly had no intent to profit, LaMacchia was charged with conspiracy under wire fraud rulings instead of copyright infringement. In December of that year, the case was thrown out and his name cleared by another judge on the basis that software piracy cannot be prosecuted under wire fraud legalities. At the same time,

> The court agreed with the Dowling Court's statement that criminal penalties for copyright infringement should be limited to situations that, in Congress's view, require the deterrence that criminal sanctions provide. The LaMacchia court also expressed concern that the government's view of the case would "serve to criminalize the conduct of ... the myriad of home computer users who succumb to the temptation to copy even a single software program for private use" (1714).

And so it has. But growing fears were mounting against what came to be known as the "LaMacchia loophole" (1714). In 1995, a revision to the Copyright Act was proposed to help guard against this contingency by "expanding the definition of personal financial gain to include the 'receipt of anything of value, including the receipt of other copyrighted works'" (1714), but it was struck down in the House. In 1997, two new bills were introduced that were almost identical to each other: the Criminal Copyright Improvement Act and the NET Act, which specifically addressed copyright issues in a digital environment. The latter was signed into law that year. "It aimed to expand the reach of criminal penalties for the misappropriation of copyrighted works in two situations: first, cases in which the infringer acts without direct financial motivation, and second, cases in which the infringer acts 'with no hope of any personal gain whatsoever'" (1715). It lowered the number of copies to one or more if the dollar value exceeded $1,000 and added criminal penalties of a year in prison or a $100,000 fine, or both. Financial gain was redefined to include even the hope of receiving free software in exchange and the statute of limitations was expanded to five years. Penalties for 10 or more copyrighted works with a value of more than $2,500 were raised to three years in prison, with a maximum fine of $250,000. In 1998 came the Digital Millennium Copyright Act (DMCA). This Act was design to bring compliance for U.S. standards with the World Intellectual Property Organization Copyright Treaty and the WIPO Performances and Programs Treaty. A major new inclusion was a ban on any technology that was designed to circumvent digital locks. It also expanded the scope yet once more for criminal penalties. For willful circumvention, first offenses carried a five-year jail term and a *half-million-dollar* fine. Repeat offenders were liable for *a million dollars* and 10 years behind bars.[2] In 1998, the Copyright Extension Act extended terms of copyright for an additional 20 years.

Hence Aaron Schwartz's impossible situation in 2011 with $35 million in possible fines and serious jail time coming after just a small jump forward in time. It no longer matters whether there was evidence or not (and there was not) that he had any intention to profit from the data he was mining. In Schwartz's case, JStor (remarkably) did not even press charges – and, after such intense scrutiny and criticism, has now enabled free access to out-of-copyright materials. It is instead the university and the Massachusetts District Attorney's office that is prosecuting. The courts always seem to strike hardest at the Schwartzes and LaMacchias of the world who are outspoken opponents of restrictive copyright regulations. And Schwartz was just the leading edge of this kind of attack, since it is hard to deny that the severity of the charges against him actually reflect a backlash against the more dangerous precedent set by a similar case with a company called WikiLeaks.

WikiLeaks is an international nonprofit corporation that fights for the free circulation of information, and stands against restricted, private, secret, and classified government materials. Originally designed as a wiki, it now follows a more traditional publishing model, in part no doubt reflecting the sheer volume of materials that it controls – materials donated by journalists and anonymous news sources – and to enforce its mandate to be "an uncensorable system for untraceable mass document leaking" (WikiLeaks.org). The website reports that 1.2 million documents were added in its first year (2006), and, by 2010, 20,000 of those had been released. Culling information from anonymous sources and public databases, it "describes its founders as Chinese dissidents, journalists, mathematicians, and start-up company technologists from the United States, Taiwan, Europe, Australia, and South Africa" (WikiLeaks.org). The site is run by more than 1,200 volunteers, and Australian publisher, activist, journalist, and programmer Julian Assange. Assange is director-in-chief and a prime target of legal and less than legal actions. The controversy over WikiLeaks exploded in April 2010 when the site released footage of a friendly fire incident in Iraq that had taken place on July 12, 2007. It was the recording of American forces shooting down a group of Iraqi journalists and unarmed civilians. A flood of hundreds of thousands of documents about the war, internal U.S. state department materials, and secret documents relating to Guatanamo Bay detentions followed over the next few years. "The organisation's stated goal is to ensure that whistleblowers and journalists are not jailed for emailing sensitive or classified documents, as happened to Chinese journalist Shi Tao, who was sentenced to 10 years in 2005 after publicizing an email from Chinese officials about the anniversary of the Tiananmen Square massacre" (WikiLeaks.org). Since then, Assange has received a number of awards, including the Amnesty International Media Award, for publishing details about murders in Kenya, and was picked Reader's Choice for *TIME Magazine*'s Person of the Year in 2010. (It is not without irony that Mark Zuckerberg, the founder of Facebook – what Assange has called "the most appalling spying machine that has ever been invented" (Houston) – was *TIME*'s own hand-picked winner for the award that year instead.) Since then Assange has also been charged with a wide variety of crimes, from sexual assault and coercion to spying in the wake of an all-out assault by the American State Department and others to shut WikiLeaks down. Whether the sexual charges are real or not, it seems surprising, if not suspect, that an alleged assault happened contemporaneously with all of the other accusations. WikiLeaks' archives have since been published by the organization in many different locations to ensure their survival after various governments tried to silence them and shut them down. A series of mishaps, probably accidental, led to confidential data leaking out of the archive itself through three sources: the interventions of WikiLeaks itself, *Guardian* reporter David Leigh, and Daniel Domscheit-Berg. Domscheit-Berg, Assange's competitor, wanted to jump on the WikiLeaks bandwagon and is believed to have stolen a copy of the encrypted WikiLeaks files. While anonymity is always preserved in the materials it releases, WikiLeaks' encrypted archive contains much confidential information. The theft would apparently not have been a problem if investigative reporter Leigh had not published the password to the file a year earlier. As a result, all kinds of information about agents in the field, journalists, Afghan informants, and many other people whose lives could be at risk as a result was inadvertently released. While lawsuits are still pending, Assange fights extradition, the American bank blockage of WikiLeaks assets continue, and donations can no longer be made to them through the PayPal account (suspended on account of 'illegal activity') or to Visa or Mastercard (suspended until Datacell can prove that WikiLeaks is not engaging in "illegitimate transactions to fund the WikiLeaks site" (Datacell)). Despite this, WikiLeaks continues to do the job it set itself. In September 2011, for instance, it released a U.S. State Department cable which showed that the Motion Picture Association of America was behind a recent drive to increase fines and penalties against Internet Service Providers (ISPs) *in Australia* where it has no juridiction. The Australian Federation Against Copyright Theft (AFACT) contracted out the monitoring of bit torrent sites in 2008, and then delivered a "telephone directory-sized" list of infringers demanding action to iiNet (Lasar). iiNet could not proceed without a court order and the court upheld that decision (although the case is being appealed). WikiLeaks revealed that this was not an Australian initiative at all, but that in fact the pressure to criminalize was coming from Hollywood. Matthew Lasar states, "This latest revelation revives a question raised by earlier WikiLeaks cables. How many Commonwealth member nation anti-piracy initiatives are essentially a creation of US content rightsholders, or the US State Department, or a combination of both parties?" A May 2009 WikiLeaks cable demonstrated that the U.S.A. was helping finance far stricter law reform in New Zealand as well (Lasar). The more WikiLeaks has been undermined by recent financial limits however, the more its model spreads. Copycat sites propagating the Leaks principle are springing up, including Domscheit-Berg's OpenLeaks, HackerLeaks, and ArtLeaks.

Everything digital is a copy of a copy, but it is only with the digital age that the perfect copy surfaces; that is, a copy with no originals. Historically making copies of things, which was one reason why artists' likenesses were so highly prized, was difficult. Photography altered that and, since digital technologies have made the whole history of human materials available, the new artistry now lies in how we remix pre-existing materials. In a digital age, more than any other time before us, we live in a remix culture. Remixing has always been with us. It used to consist of taking a work of art or, say, a novel and re-imagining that work to re-express it, like Peter Jackson's homage to Meriam C. Cooper's *King Kong* or Alice Randall's remix *The Wind Done Gone*, a retelling of *Gone with the Wind* from the perspective of slaves. You might think there is nothing unusual about these borrowings. Most if not all human cultures have developed by building on the foundations of older ones:

> Ancient Rome remixed Ancient Greece; Renaissance remixed antiquity; nineteenth century European architecture remixed many historical periods including the Renaissance; and today graphic and fashion designers remix together numerous historical and local cultural forms, from Japanese Manga to traditional Indian clothing (Manovich, 2005).

But within digital culture, this notion of remixability moves to a whole new level. Digital copying is not just the transposition or re-creation of an original in another medium, but the literal, indistinguishable, identical copy of another digital original. This is because every blip, every click, every algorithmic execution within the digital machine involves the function of 'copying' be it saving to hard drive or opening a stored file in Random Access Memory, or downloading an MP3 file from the Web. A moot point you may think, unless you happen to be a copyright lawyer.

Copyright law, born in Seventeenth-century England, was designed to protect property from encroachments on a creator's right to profit from it and an owner's right to use it. This made sense when what we were talking about was a material object that could only be used for one purpose at one time. It meant that you could not republish my book without my permission and it meant that you could sell or give away a book that you had already read. Up until 1909, the law in the United States regulated only publishing. In 1909, however, one word was changed: "'copying' was substituted where it had previously said 'publishing.' Technically, [according to Lawrence Lessig] the change was an accident, a mistake … But that 'accident' had enormous implications as the technology for copying things changed" (Marshall, 2004). This is now the driving force behind a massive move by the American and other legal establishments to redefine remixing and therefore any kind of reproduction as *piracy*. This might include things like rereading an ebook that you have already purchased or rewatching your pay-per-view movie after the allotted time. It will definitely include that Harry Potter-inspired sequel you are writing.

I do not mean to imply that bona fide piracy is not a problem or a multi-billion-dollar industry. It is. In China, Vietnam, Indonesia, and Russia, for instance, at a conservative estimate 86 percent to 90 percent of all computer software is pirated (Associated Press, "China's Piracy"; da Silva). It is believed that 95 percent of all music CDs in China are pirate copies (Mason, 2008, 159); but think about the fact that the lowest piracy rate in the world – which is the United States – stands at 21 percent (Stangrunner, 2006): 21 percent! That is astronomical. If more than one-fifth of all the digital copies that are being made are being declared pirate copies (and that is in an affluent nation), then what does that say about how skewed this definition of digital piracy might be?[3] Might there be systemic reasons, for instance, as to why it is not viable for consumers to purchase legitimate copies in particular countries or situations in the same way that Prohibition was not viable? Let us think too about how the application of a recycling aesthetic or an open source model could redefine these so-called acts of piracy as legitimate use within a commodity culture. In this book I will explore some of the realities of creative expression – a.k.a. piracy – in non-Western nations, with a particular focus in the final section on Pakistan and China. Creativity as we have known it is dead, and from its ashes has arisen Remix Culture, a free culture movement encompassing multiple modes of expression that will not be stopped, no matter how draconian the laws against it. But there are also other models of creative practice that point the way to new and ever more inventive remix methods and applications. These instances of 'digital anthropophagy' and 'productive mistranslation' demonstrate highly inventive approaches for thinking about creative practice and repurposing existing materials and ideas outside of Western models.

Piracy as the law defines it is prevalent to be sure, but it is also a much more complex issue on the global stage than copyright enforcers want us to believe. Media critic Brian Larkin has observed, for instance, that "For the vast majority of Nigerians, Indians, or Egyptians … the array of global media is only available through the mechanism of piracy; piracy is thus the default infrastructure through which nearly all foreign media flow" (Larkin, 2007, 78). In China, filmmaker Hao Wu is hungry to see films that might help him improve his art. The Communist government allows only 20 sanctioned overseas blockbusters to be released at Chinese theaters per year. Some legitimate DVDs are available for purchase for about $20 each, but the average urban Chinese makes only the equivalent of $1,000 per year; in rural areas, the income is less than one-third of that. Bootleg copies cost about 50 cents each. Hao says that pirate copies are the "only means of unfettered information access" available to him. In 2007, when TIME-Warner lowered its price on DVDs to about $1.85 each, he was willing to pay more. What he got was much less. Time-Warner had deleted all of the extra features, *including the language selection option*; the film was also edited for content removing a bare shoulder – a feature the government had declared inappropriate nudity. That is the DVD market. Software is similar. With the cost of the Windows operating system set at the same price in China as in North America, what are the chances people are going to lay out $300 when the pirate version may be had for a few dollars? – and that doesn't take into account the fact that Linux, the officially sanctioned operating system of the People's Republic of China, is free. Microsoft acknowledged its unrealistic pricing in 2008 and in a landmark move announced plans to lower its prices. However, it has also added a built-in legitimacy check, so if your version is not authorized Microsoft will shut your computer down.

China has masterfully managed the transition to the technological age over the past 20 years. Relatively poor in intellectual property as a result of a half century of enforced conformism, it uses its immense wealth in human labor to attract technology into its midst. What is known as "[t]echnology transference" – that is, building high-tech devices for the rest of the world – has catapulted the nation into the world market, but has left it technologically-dependent' (Wang, 2007). In the short term. That tide is now changing. "The importation of incorporeal ideas and ... cultural software ... always spurs the creation of lots of physical stuff with potential economic value [or] ... cultural hardware .... The synergistic effect on the economy becomes enormous" (Moser, 2006). This process started to reverse in 2002 when ex-patriots came back home to China to escape the increasingly hostile post-9/11 Western world. They started a move to clone the social media engines of the West:

> A Silicon Dragon tech economy began in 2002 with Chinese returnees – so-called sea turtles who came home to lay their eggs – who cloned Google, YouTube, and Amazon, grabbed Sand Hill Road money, and scored on NASDAQ and the NYSE. Today, homegrown Chinese entrepreneurs are snapping up venture capital from Chinese currency funds for even more clones – Beijing techie Wang Xing alone has cloned Facebook, Twitter, and a Chinese GroupOn (Fannin, 2011).

In 2006 China's leadership announced a new policy to "create an innovation-oriented country" (Wang, 2007), which seems to be working to foster new and more original creative enterprises. In September 2011, *Forbes Magazine* ran an article entitled "From Made in China to Invented in China: 'Chinov8!'," which focused on a conference being held by the Silicon Dragon, the cluster of Silicon Valley innovators sprawling outward from Beijing and Shanghai and elsewhere in the Middle Kingdom:

> The needle is gradually moving from "made in China" to "invented in China." Micro-innovations tweaked for the local culture are cropping up more often. Sina's Weibo, a hybrid Twitter-Facebook, layered in video and photo sharing before Twitter did. The long-awaited promise of disruptive technology from China is coming, too, symbolized by China's climb to fourth place worldwide for new patent applications (Fannin, 2011).

This will be an uphill battle after decades of policy to the contrary, but the digital piracy industry has laid the foundation for a whole new remix of Chinese and other cultures. We have already seen the effects of this in India where "*the cassette revolution* of the 1980s transformed popular music culture in the first phase of globalization, [which was then] largely nonlegal, and in doing so effectively broke the stronghold of the large music companies by introducing new artists and expanding the market for low-cost cassettes" (Sundaram, 2007, 55). Japan has for too long been a master of remixing Western culture and things have come full circle with the arrival of the film *Sukiyaki* (as in noodle) *Western Django* (Miike, 2007), a Japanese samurai spaghetti western that remixes the Italian remix of the American western. In Russia, the biggest blockbuster of the post-Soviet film industry is a pair of films called *Night Watch* (Bekmambetov, 2004) and *Day Watch* (Bekmambetov, 2006). These genre vampire films remix the *Lord of the Rings Trilogy, The Matrix*, and *Buffy the Vampire Slayer* with the unique flavor of Russian bureaucracy.

The foremost Russian movie translator Dmitri Puchnor (who is known as The Goblin, and who did the official translation for films like *Pulp Fiction* and the *Kill Bills*) has acquired considerable infamy for an unofficial remix of the first two films of Peter Jackson's LOTR with new dialogue in the language of Russian slang and profanity. The hobbits and other good guys are cast as "bumbling Russian cops, and the 'bad' Orcs" as Russian gangsters" (*Observer*). "Frodo Baggins is renamed Frodo Sumkin (a derivative from the Russian word sumka, or bag). The Ranger, Aragorn, is called Agronom (Russian for farm worker)" (*Observer*). A reviewer in the *Observer* says, "Gandalf spends much of the film trying to impress others with his in-depth knowledge of Karl Marx, and Frodo is cursed with the filthy tongue of a Russian criminal." Originally created as a spoof for The Goblin's friends, when it was posted on the Web it became a runaway Russian hit, now selling at about $10 a piece in Russia – to the infuriation of copyright lawyers (*Observer*). The Goblin's work straddles the line between legal and illegal. Satirical material is technically legal, but because Puchnor's translation is a satire on Russian sensibilities and transposes it into the foul language of Russian low lifes, it could be argued that the changes are part of the cultural component of an unauthorized translation.

New industries and a new generation of entrepreneurs spring up around bootlegged property. Ironically enough, it was the arrival of tons of American garbage that started the digital cultural revolution in China in the mid-1990s. So-called sawcut CDs are overstock or discontinued discs that the industry has slated for demolition. Immediately recognizable by the inch-long notch in their side, they remain playable because CDs read from the inside to the out. These were shipped by the boatload to China to be recycled, but were far more profitably turned over for about $1 a piece on the black market. In the wake of this:

digital technology hit China like an atomic bomb. Virtually overnight, waves of pirated CDs, software, computer games, and VCDs became available through underground bootleg channels. The effect was not merely the opening of a spigot; China was suddenly inundated with a veritable tsunami of foreign … intellectual products. […Soon] Pirated movies could be purchased almost the same week they were released overseas, making

Chinese movie fans instantly conversant with current film trends (Moser, 2006).

The rise of independent rock bands, Mandarin hip-hop and the flourishing Chinese film industry are a result of stronger and stronger links with the ideas of other music and film cultures worldwide. But remember: if you own a book, you can resell it. If you sell a CD or DVD you found in a boatload of American garbage, that is piracy. That is the politics of the digital. Gone are the days when you could simply sign a name on a urinal and display it in a gallery.

Perhaps it is no accident though that the American form of hip-hop and turntablism, the use of the playback medium as an instrument, also grew out of free materials, materials that were liberated during the July 13, 1977 blackout riots in New York. "It is that day," Boon says,

> which is credited… as being the moment of hip-hop's tipping point, where the technologies required for MC-ing and DJ-ing (turntables, microphones, and speakers), formerly available only to a small number of crews, were suddenly in the hands of just about anyone who wanted them. This free access facilitated hip-hop's full emergence as a culture (Boon, 2010, 43).

Another artist, Christian Marclay, a British sound technical, sculptor, remix, and performance artist, also used the turntable and vinyl records themselves as instruments for the creation of music or, more correctly, what he dubs "the sounds that people don't want: the pops and the clicks and the scratches" (quoted in gmooney, 2006) before the arrival of hop-hop in the 1970s. He then incorporates those sounds as part of the rhythm and the music of his compositions. Marclay got his start, with the permission of punk rock, by playing media players to create noise that he collected for performances. From there he graduated to cutting and inlaying vinyl records together like slices of pie or checkerboards to create disconnected sounds that looped or repeated at intervals. The effects, from a Twenty-first-century perspective, sound like a lightning bolt from the future; they sound like the effects of digital sampling, which reuses pre-existing works to create new works from their recognizable parts. Today, after many reforms of international copyright law later, what Marclay did in the 1970s would be labeled piracy, and his present work with video, where he unites sound with video clips as a gallery installation, won him the Golden Lion at the Venice Bienniale in 2011. Ironically, if he tried to distribute or sell "The Clock" it would either land him in jail or cost him millions in rights (if all of the owners of the clips could actually be tracked down).

Another conceptual artist, Chinese-Dutch Ni Haifeng, recycles primary materials that are sent back to China for recycling or disposal and reuses those materials to create trademark Dutch objects out of them. After Ni relocated to Amsterdam, he was shocked by how many of the everyday objects he used in his new city were actually made in China. Intrigued by how globalization was creating a displaced world of objects that had been manufactured in his homeland, Ni started to explore this immersive virtual China in his art. He says,

> I began to incorporate in my work the process of production during 2004 to 2005, in a project called *Of the Departure and the Arrival*, in which I commissioned a factory in Jingdezhen to produce a large amount of porcelain copies of various discarded Western objects. The whole project is a mockery of both colonial trade and global commerce, and the process mimics the inner mechanism of global production and consumption. Production, here takes on two qualities, that of the regular handcraft and the artistic. The former leads us to interrogating the labor condition of traditional handcraftsmanship and the latter creates a social dimension of collective authorship in the production of art (Haifeng and Yao, 2011, 41).

By shipping samples of Dutch handiwork back to China from the Netherlands, Ni creates a kind of collective authorship. Delft locals donate copies of their ceramics for Ni to work with. In Jingdezhen, he has replicas of the city of Delft's famous *handcrafted* ceramics manufactured to his precise specifications. The highly valued blue and white originals are used to make molds so that the copies are perfect in every way. Ni then ships the worthless but perfect Defltware counterfeits by sea back to Europe where they are carefully exhibited in installations, with many copies still in their original shipping crates. In a later work, *Return of the Shreds* (2007), Ni again explores trade and global manufacturing. In a reversal of the usual trade trajectories, Ni shipped several tons of discarded fabric from a Chinese manufacturing plant in Zhejiang to a contemporary art museum, Stedelijk Museum De Lakenhal in Leiden, that was a converted blanket factory. The manufacture of the blankets has undoubtedly now been relocated to China. The art itself was a wall hanging woven of detritus – offcuts and discarded scraps – from the Chinese factory that was displayed next to an assortment of other kind of imports and manufacturing offcuts, including broken porcelain, tea-leaves, and spice. The origins of this garbage were made manifest in the documentation of global trafficking's shipping, customs, and excise documentation. That the final recycled product had no purpose underlines the fact that the real work of art lies in its *documentation*. This is a topic I will return to in the next chapter in reference to the conceptual nature of digital art.

To return to bona fide bootlegging (as it were), the coming of the World Wide Web to less privileged countries has accelerated pirating trends to light speed. "Digital information is inherently 'leaky' – electronically transmittable and reproducible and highly contagious – and digital products are cheaply produced in mass quantities, easily transportable, and extremely profitable" (Moser, 2006). Theorist Ravi Sundaram sees pirate culture as "a *just-in-time* culture. The copy arrives on your cable network the weekend the film is released, and the music versions of popular numbers follow almost immediately" (2007, 57). In Russia, where organized crime controls digital piracy, those gangs own the same high-tech equipment that the Hollywood studios use to convert 35mm film to digital format (AP, 2006). These high-quality copies generally hit the street three to five days after a film opens at theaters (AP, ibid.).

Piracy is so incredibly prevalent in part because of the ease with which it may be practiced. Anyone with a $1000 computer can go into business – sort of like setting up a still in your kitchen in the days of Prohibition. In China, to set up a currency counterfeiting operation, it will cost $1000 to generate $5000 profit. Heroin trafficking would return $50,000 with the same investment. Yet a $1000 investment in pirated pharmaceuticals or digital media will net half a million (Iyengar, 2004). There is also a real hunger for this material. In China, the government does little to discourage piracy because it is a powerful opiate for the masses to be sure, but also because it is a multi-billion-dollar industry. While the West cries foul and claims it is losing $50 billion in DVD sales per year, most of these films would never have been admitted to countries like China legally (and Hollywood does not even care enough about small markets, like Nigeria, to release its films there in the first place). The pirated products have in fact now opened the door to legitimate sales and the legal distribution of films, as with the recent release by Time-Warner in China.

Unfortunately, much ground continues to be lost, especially as fines and jail sentences continue to escalate. In Karachi at a large piracy mall in the main market, there is a union which helps protect those who produce pirated materials from prosecution, fines, and even jail time. That the noose gets ever tighter is revealed by the Department of Homeland Security's revelation that it now routinely copies from "books, documents, and the data on laptops and other electronic devices without suspecting a traveler of wrongdoing" (Nakashima, 2008). The Association of American University Professors warns that:

> There is little or no information about how information that is copied and kept by Homeland Security will be kept secure, leading to concerns about the protection of original research. This [also] extends to projects that may have patents pending or are in an otherwise precarious stage of development (Nelson, 2008).

If corporate interests were twitchy about copyright before, it is anyone's guess what this will do to change it. The AAUP goes on to say that "It is unlikely that the Customs and Border Protection agents conducting such searches at the border would have the specialized knowledge to determine whether or not certain types of data, particularly in areas of science such as engineering or biochemistry, pose a genuine threat" (AAUP).[4] Fair Use, in turn, has virtually ceased to exist as a result of changes to copyright law everywhere, including a devastating agreement by YouTube with Viacom, which I will discuss at length in Part Two.

China is at the center of this "counterfeiting epidemic" which accounts for 7 percent of all global commerce (Wang, 2007), but this also marks a whole new Chinese renaissance of creative culture. Initially, many of their new works were derivative – baby steps in the process of apprenticeship into original input in the creative arts. Chinese rock star Kaiser Kuo observes that bootleg culture has:

> penetrated deep into the interior, opening a window into the Western world otherwise inaccessible to the Chinese hinterland. The great boom in the Chinese film industry, the explosion of rock music talent coming out of Beijing and other Chinese cities, even much of Mainland China's Internet revolution – all this has in large part been made possible by the piracy phenomenon (Kuo, 2006)

In direct contradiction of China's xenophobic policies in the past, the Chinese government publicly dubs this Western invasion of ideas "spiritual pollution," but despite public posturing privately turns a blind eye. This is because not only is there big money in free culture, but it is an essential foundation for building a society rich in creative property. More important for artists, they do not merely copy material, but access is allowing for a kind of decontextualized cultural translation as a century of (primarily) American culture becomes available all at once. The same process occurred in the 1990s in the Chinese art world with electrifying effects for the international art scene.

With the encroachment of ever more repressive copyright and patent laws in the West, we are starting to see the early effects of these ever more restrictive policies. Michael Heller, a leading authority on property law, has explored in his book *The Gridlock Economy* how too much ownership of creative property is stifling innovation, research, and development. Gridlock in our airwaves, for instance, led Heller to call it "the most underused natural resource in America" (2008, xiii). Ninety percent of broad spectrum bandwidth is dead air – or was until November 4, 2008 when the Federal Communications Commission repealed this practice (Heller, 2008, xiii; Howard, 2008). Japan and Korea have far superior coverage in terms of bandwidth because in America there have been too many owners fighting over the right to control it. In 1996, Congress voted to break up Bell Telephone monopolies thereby forcing them to unbundle the network elements so that up-and-coming enterprises could buy up what they needed to start new services (Heller, 2008, 104); (in 2011, a merger between AT&T and T-Mobile came dangerously close to passing, which would have effectively returned AT&T back to the state of Bell's 1980s monopoly.) Instead phone and cable companies fought over the pieces like rabid dogs, and the end result was that the price of broadband has remained high. Where other countries have approached this space as public property, its privatization has hobbled the American public's ability to access broadband for years. This new ruling "allows innovators to develop new technologies that will bring Internet service to millions of Americans in underserved communities" (Howard, 2008). Competition and innovation should spark new growth in this area at a time when the economy is otherwise on "a downward spiral" (Howard, 2008). That remains to be seen. On July 1, 2010, Finland became the first country to make broadband access a legal right; Spain is not far behind (Catacchio, 2010).

Gridlock in academic research in medicine and pharmaceuticals[5] is far more disturbing. An assessment by sociologist John Walsh in 2003 discovered that researchers routinely acted illegally, flouted patent laws, and used patented research tools without a license. He said, "University researchers have a reputation for routinely ignoring IP [intellectual property] rights in the course of their research" (Heller, 2008, 66). Part of the problem with this is the expense and difficulties in following the law:

> to investigate ownership of the intellectual property used in a single campus lab, the University of Iowa had to contact seventy-one different entities and spend tens of thousands of dollars in background checks …. Ultimately, a property rights regime that turns researchers into patent pirates is corrosive (Heller, 2008, 66).

A second survey by Walsh found that rather than do battle with these laws researchers regularly "abandon research that involves infringing activities" (Heller, 2008, 66–7). They "also deploy a range of ad hoc solutions: they … work around troublesome patents, move research out of the United States to offshore labs, pirate intellectual property, or even litigate" (Heller, 2008, 67). Clearly this is not a healthy climate in which to conduct innovative work. One of the projects that has been blocked by this patent gridlock is a successful cure for Alzheimer's Disease. Heller reports that patent holders have actively prevented it from coming to market.

This gridlock is apparent in the arts as well as the sciences. A landmark case was the documentary *Eyes on the Prize*. Called "the principal film account of the most important American social justice movement of the 20th century," this 14-hour documentary brought together interviews with hundreds of people who knew Dr. Martin Luther King, and included "video footage from 82 archives … 275 still photographs from 93 archives, and some 120 songs" (Heller, 2008, 10). The original licensing for the film expired shortly after the film was first shown. Rights to the individual media clips changed hands. The original permission did not include rebroadcast or DVD rights and so the film was shelved for almost 20 years. With the new complexities of copyright law the film could not even be made today, and it took a $600,000 donation from the Ford Foundation and many hundreds of thousands more from private donors to be able to buy up the rights to rerelease the film in 2006 (Heller, 2008, 10–11).

As a result of the complexities involved in this project and others, documentary filmmakers banded together to try to settle issues of fair use. Frequently filmmakers were being motivated by the fear factor and as a result were paying for rights that should have been covered by fair use: the appearance of a logo or a song in the background of a film clip, for instance. Fair use becomes thornier and thornier ground as copyright keeps being extended longer and longer; "imping[ing] more and more directly on creative practices," its importance is growing (Center for Social Media, 2008). Fair use, filmmakers argued, "is a right, not a mere privilege. In fact, as the Supreme Court has pointed out, fair use helps reconcile copyright law with the First Amendment" (Center for Social Media). Far too many filmmakers were shying away from making films like *Eyes on the Prize* owing to fears of legal entanglement. Furthermore, "they routinely altered the reality of the localities where they captured images, and they also changed both sound and picture after the fact to protect themselves" (Aufderheide, 2007, 29). Confusion reigned, with many filmmakers and their lawyers confusing trademark law with copyright law and others assuming that their projects had to be nonprofit for these rules to apply. Assessing past rulings in the courts, they came up with four principles for fair use in documentary films:

1 Copyrighted material may be used as the object of social, political, or cultural critique.

2 The quotation of copyrighted works of popular culture may be used to illustrate an argument or point.

3 Copyrighted material shown in the process of filming something else is allowed.

4 Using copyrighted material in a historical sequence is acceptable.

Together, these four rules are being combined with the two criteria the courts use to assess works for infringement: (1) Did the work transform material rather than simply parrot it?; (2) "Was the amount and nature of the material … appropriate in light of the original work and of the use?" (Center for Social Media).

This proactive stance by documentarians is starting to open doors for other fields in their assessment of fair use practices. Lawrence Lessig has observed that "Fair use has long been misunderstood and misrepresented as a minor feature of copyright law, difficult and even dangerous to use. It has been disparaged even by advocates of the more flexible and balanced copyright law" (quoted in Aufderheide, 2007, 26). The Modern Language Association has devoted a whole chapter to "Legal Issues in Scholarly Publishing" in the third edition of its *MLA Style Manual and Guide to Scholarly Publishing* (2008). Influenced by the documentary filmmakers' rulings, the MLA states that "The purpose of fair use is to advance creativity and public knowledge" (50). The guide reiterates that with regard to "purpose and character of the use" courts are more lenient when the uses are "criticism, comment, new reporting, teaching … scholarship, or research and to uses that are for nonprofit educational rather than commercial purposes" (51). As long as the use is transformative, that is to say that it adds to the existing work, then we are on firmer ground.

There is still a lot of misinformation about fair use floating out there in the cybersphere. If you buy a DVD, it is illegal for you to copy it to your laptop; it is also unlawful to listen to your iTunes song on a non-Apple MP3 player. But while there has been an aggressive ad campaign against illegal downloading of digital files, downloading is not actually illegal in the United States. Copying, however, is, as I have already mentioned. In the process of downloading a file, it is copied to your hard drive.

French artists Philippe Parreno and Pierre Huyghe's anime work, *No Ghost Just A Shell* (1999–2002), presents another alternative – and offers a built-in critique of the copyright system. A riff on the phenomenally successful manga, anime, video game franchise *Ghost in the Shell,* Huyghe and Parreno translate a stock anime character into a new model for the creation of authorship. Playing on the cyberpunk cyborg character created by Shirow Masamune (1989) in the original manga comic and fleshed out and brilliantly translated by Mamoru Oshii (1995 and 2004), Parreno and Huyghe purchased a stock manga character template from a Japanese ready-made agency. Fully formed characters are extremely expensive and come with rich character attributes; versatility is a costly trait that allows it a longer lifespan through multiple media incarnations. By contrast to most manga characters, the character Annlee is unfinished and her eye sockets are empty. What the duo sought to foster in her, instead of a mere animation, was a level of inter-activity and *conversation* that encouraged people to develop the character for themselves. They created a film each, *Anywhere: Out of the World* (2000) and *Two Minutes Out of Time* (2000), "in which Annlee reflects on her absurd and tragic situation" (Paul 111). She has "no history and no life; she is fictional shell with a copyright, waiting to be filled with a story" (Paul, 2008, 111). Taking this recycled background character from a Japanese cartoon, the character steps into the role of commentator on her own situation: "'While waiting to be dropped into a story,' the girl says about herself, 'she has been diverted from a fictional existence and has become a deviant sign'" (quoted in Rush, 2007, 172). Huyghe and Parreno then expanded the concept of authorship by not keeping her as a discrete creation, but by giving her to other artists to reimagine Annlee as a part of their own repertoire. The artists who did included Liam Gillick, Dominique Gonzalez-Forster, François Curlet, Pierre Joseph, Richard Phillips, Joe Scanlan, Rirkrit Tiravanija and Anna-Léna Vaney – who provided expensive video and extensive tech support for each of the other artists – to name a few. (Other artists involved included Douglas Gordon and Catherine Deneuve.) Each of these artists conceptualized her in "the French surrealist tradition of the 'exquisite corpse'" (Tanner, 2002), and filled her shell with ideas and images including video animation, books, posters, paintings, neon, and sculpture. After a show at San Francisco MOMA, Annlee was retired with her copyright being signed back to herself. The original story had been about a female cyborg Major named Motoko Kusanagi in the Public Security Section 9 counter-terrorism unit in a futuristic Japan. Specializing in technology-infused crimes, Motoko is the CO. Empty-eyed Annlee, by contrast, awaits the interactor to endow her with a character. The resulting projects were not sequential or connected in any way apart from her image, but the character creates its own Web of connections or Internet of things. "Each of the projects realized with 'Annlee' is a 'chapter in the history of a sign', and has a 'life' in the context of the individual artists' activities and within the joint project" (Zürich Catalog, 2002). With each inception of Annlee, her story and our opinion of her changes.

> Philippe Parenno and Pierre Huyghe, the 'authors' of the exhibition have tried instead to create an exhibition complex that is again intended to give access to central aspects of the project. Both artists, in their own projects and their collaborations, are not so much interested in the final artistic product as in creating a set of relationships and processes on the way from production via distribution to reception. Shifting perception from representations of objects to interpreting their forms and defects become central (Zürich Catalog, 2002).

Endowing her in such a way makes her a *conceptual* art object ripe for interpretation that continues to be of interest. She does not exist until an artist/author begins to write/draw her. The original *Ghost in the Shell* character had floated around in a world that is "made borderless by the Net":

> The secret agent of the future is a non-human entity without physical body that freely travels the information superhighways – until it decides that it is a life form in its own right and requests physical existence. While we may not yet be ghosts in a shell, the futuristic scenario of Shirow's *manga* and the issues addressed in Parreno's and Huyghe's work point to several of the key questions that have been raised by the Internet and its use as an art medium (Paul, 2008, 111).

Annlee has no ghost, no interior life – she is only shell. It is through creative intervention that this copyrighted character acquires the digital equivalent of a 'soul'. It is only after all of the above-mentioned artists have imagined possible lives, goals, and personalities for her that she acquires a kind of digital personhood, and gains control of her own previously shared copyright.

Lessig's Creative Commons has gone a long way to work around the gridlock monolithic interpretations of copyright pose. By licensing our own works for alternative uses, we can open up creative gates that would have otherwise been closed. Composers have been leading the way with remix initiatives led by Gilberto Gil, Brian Eno, Danger Mouse, and others. Creative Commons has changed the rules of creation, as the millions of licenses issued attest to. Repositories like iTunes U where some universities now submit all of their lectures as podcasts, and recent initiatives by universities like Harvard and MIT to make all of their scholars' works available through similar kinds of licensing, attest to the importance of new initiatives that keep creativity alive in and for the future. While the Chinese model of only periodically enforcing piracy just may make the case as to how important creativity is to the future of a people, helping to rewrite the future through approaches such as Creative Commons licensing may spare us the dire fallout of a new Prohibition.

This book will explore the politics of authorship and the act of creation, its history, and some possible future trajectories, by making explicit its complex facets and media-specific issues. Part One, "The aesthetics of appropriation," will explore the death of creativity and the rise of three new aesthetic practices in new media in an age of perfect copies. The first is interruption (which includes a process of stoppage and repetition), and marks a shift from the old creative model's focus on *content* to a *process-oriented* work of art that opens an image to create a space for critical engagement. The second, disturbance, which brings together an action with an event, is a new kind of aesthetic activism – artivism or tactical media – for the socially networked age. Looking at the psychic and social consequences of technology and aesthetics, this section explores the importance of performance in an age of consumer culture for the creation of tactical media as an agent of new political practices and events. The third practice is that of capture/leakage, which is a meeting of performance with documentation. In an age of information overload, everything is potentially subject to capture and to surveillance. At the same time, once the data exist they are always already potentially leaky; the more tightly information is controlled, paradoxically, the more likely it is to be released out into the data environment. In the spirit of protocol and data, documentation is coming to the fore as a new art form in its own right.

Part Two, "Authorship," explores new works and practices derived from sampling, mashups, remakes, datastreaming, the application of Situationist philosophies, and other approaches that engender a different kind of creative engagement for readers. These new emergent paradigms and altered modalities are the native space-time of Twenty-first-century creative culture and digital aesthetics. Part Two will explore how real time and the spatial and tactile nature of the new media are transforming the role of the spectator, birthing new generative forms and privileging space over time in born-digital narratives.

Part Three, " Creative cannibalism and digital anthropophagy", will examine the practice of cultural cannibalism (the patenting of things 'primitive' from cultural traditions to tribal remedies) and demonstrate how digital anthropophagy and new Asian "productive mistranslations" are producing new art and practices as a form of cultural critique that just might bite consumer culture on the ass.

PART ONE

The aesthetics of appropriation

Creativity is dead

Creativity as we have known it in modern times is dead. The creative process, however, is alive and well. The art of appropriation and appropriation art are part of a long-standing critique of representation and are cornerstones of the creative process. All work builds on its predecessors. Appropriation art and its practice has gone by many names: collage, bricolage, ready-made, found-footage, merz, remix, mashup, sampling, homage, intertext, paratext, postproduction, and on and on. This process of combining old and new materials is not something unique to visual media either. From William Shakespeare's free-flying adaptations to Herman Melville's mashups to Jorge Luis Borges' "Pierre Menard, The Author of Quixote," the tale of a perfect re-creation of *Don Quixote* that bore no resemblance to the original, to Helene Hegemann's remixed social media content novel *Aoxlatl Roadkill*, it was another writer, Oscar Wilde, who famously summed up the process as "talent borrows, genius steals." Movies also follow this template. Some decades ago Dorothy Parker reflected on the shallow nature of the Hollywood dream factory by quipping that "the only ism Hollywood believes in is plagiarism." As I have previously stated, the most successful corporate media empire, Disney, is built on a foundation of reworked material. But times and the law have changed. Ironically, as Disney Corporation fought so hard to prevent Mickey Mouse from going out of copyright by continually having the duration of copyright extended (hence the name, The Mickey Mouse Act), it also kept at bay its own access to Walt Disney's own animated character, *Oswald, The Lucky Rabbit.* Oswald had initially been far more successful than Mickey Mouse was, but when Disney asked his employers for more money and more staff to make more Oswald shorts they shut down his studio and stonewalled him from using the character again because they owned the rights. Oswald became the model for all later animated rabbits, but Walt Disney literally went back to the drawing-board and started drawing mice – or two particular mice at least – instead. Disney Corporation has now reintroduced Oswald back into its empire (in the game *Epic Mickey*), but while Oswald is ostensibly public domain now, I guarantee you that if anyone except Disney tries to use his image for their own ends Disney will prosecute them.

It was with the advent of video and video recorders that what we think of as the contemporary remix was born, although the practice as a creative technique is clearly as old as both music and storytelling. We think sonically in phrases and visually in images, metaphors, and symbols, and more and more we 'talk' and 'talk back' in this mode too. Video – or digital video more precisely – has become a vernacular and, being a vernacular, it is never single-channeled. It is participatory and complex. While this language of digital imagery had its roots in time-based cinematic mode (which remains consistent at 24 or 30 frames per second), like other code-based modes it is spatial rather that visual. Its layers are a surface graphic and often metaphorical interface (think desktop) that the user interacts with and a skeletal structure made up of code or scripting languages. In fact, Christine Paul deems the digital image "not visual" at all on account of those layers of bitmapped and spatialized attributes (Paul, 2008, 68). Digital images are dynamic (even still ones) and have layers and abilities analog images never dreamt of. Digital video is indexable, searchable, networked, embedded, and available by subscription. Out of that fertile participatory ground the remix springs. Visual blogs, veracular video, vlogs, YouTube, Vimeo, and UStream are becoming the way in which a large percentage of ordinary people publish and distribute to their similarly networked peers. Of the great revolutions in communication technology, where the alphabet (and ultimately movable type) supplanted orality and shifted our primary sense to the eye, the telegraph and electric technologies supplanted print, retribalizing us (McLuhan would say) and restoring the balance of our senses; now, binary code, and networked and distributed technologies have supplanted the electric technologies and are extending our consciousness. Code may form the underlying bones of our visual and distributed cultural nodes, but culture is always participatory and interactive. The creative act itself, as inflected and informed by digital technologies, is undergoing a metamorphosis. The older media and modes of discourse – print, painting, cinema, television, etc. – are still with us and remediate new technologies' looks and methods, if not their highly political intent. Medium specificity, however, is gradually being pushed aside or rendered irrelevant as all media are becoming digitized. According to Alan Kaye, the designer of the first graphic user interface, the computer is the first "meta-medium" and, as such, it emulates all other media. Emulating, however, is not the same as being, and digging deeper reveals the very profound differences that exist between the surface similarities of traditional media and the digital. Way back in 1962, Marshall McLuhan wrote,

> The next medium, whatever it is – it may be the extension of consciousness – will include television as its content, not as its environment. A computer as a research and communication instrument could enhance retrieval, obsolesce mass library organisation, retrieve the individual's encyclopedic function and flip it into a private line to speedily tailored data of a saleable kind (McLuhan, 1995, 210).

So if the future primary media is to be located in the participatory space of networked information retrieval (and I will discuss the act of searching and Google's engine at length later in this book), then the act of searching for materials becomes a primary part of our acts of creation and consumption of digital media. What we 'own' in the act is the process not the content. Searching and the subsequent creative remixing of existing content has become the dominant mode of talking back to television, music, and networked culture.

Born-digital art forms are largely what I will be concerned with in these pages and they have their own unique grammar as a part of their creative aesthetic processes. In fact, in a digital age, we might begin to question whether *creation* even remains a useful category. In a time when we have (theoretically at least) all recorded information in the history of the world at our fingertips, creation becomes irrelevant. It becomes irrelevant because there is literally nothing new under the sun. Newness instead is now born of creative combinations that merge body, environment, and technology within generative structures – what Bill Seaman calls recombinant poetics (Seaman, 2004). In the same way that the medieval definition of genius – someone who could perform unrivaled feats of memory – was replaced by the Eighteenth-century measure of genius as someone capable of great feats of the creative imagination, so now creative genius is being replaced by acts of remixing media and ideas that produce new insights and deep critique. To reiterate then, the pinnacle acts of creative virtuosity today are in the political expression of resistance to the spectacle of consumer culture via the innovative reuse of existing materials. These new creative critiques are aesthetic and atactical (à la Giorgio Agamben),[1] and fall into three categories:

1   Interruption, which includes stoppage and repetition.

2   Disturbance, the marriage of action and event.

3   Capture/leakage, which comprises performance and documentation.

Interruption is neither visual nor verbal, but instead revisits old logics to find new answers. Like the collage, it promotes a loss of visual coherence at the same time as the logic of the verbal relinquishes hierarchy in favor of pattern recognition and parataxical architectures (Joris, 2011, 185–6). The second method, disturbance, is the only political, aesthetic alternative left for free expression in a surveillance culture in which all desire for resistance threatens to be drowned by consumerism. Combining rhetoric and aesthetics, disturbance is where action meets the critique and the event. Ephemeral, disturbances make the politics of events and actions visible. The third method, capture/leakage, is a process that is performed and later recorded via its documentation. Documentation is the closest thing to an art object that exists in remix culture because remixing is primarily process- not product-oriented. Often the domain of hacktivists, documentation is what survives after the event. Real power exists these days in the power to suppress, to be silent, to escape detection. Hacktivists seek to make the invisible visible again. The process of capture/leakage is informational and dynamic and performed in opposition to prevailing power structures. As the explication of the event or as the road-map through the newly visible information, the only remaining proof of the process is the documentation. Always surreptitious, if not sufficiently delocalized, the process of capture/leakage often results in capture and suppression of its information flows, sources, and people. All of these approaches – interruption, disturbance, and capture/leakage – are nonvisual, dynamic, interactive, and participatory. These methods of re-engagement are tied to political action, not critique. This is not the first time these modes of action have been theorized, although digital media alter both actions and their results. Poet, playwright, director, and actor Antonin Artaud wrote of the *Theatre of Cruelty*, a violent physical determination to shatter the illusions that imprison our senses. Media guru Marshall McLuhan used probes, nonbinary rhetorical constructions that were not easily unraveled. Similarly French philosopher and Situationist Guy Debord said, "All forms of expression are … reduced to self-parody…. We find ourselves confronted with both the urgent necessity and near impossibility of bringing together and carrying out a totally innovative collective action" (Debord, 1981, 55–6). I will discuss these methods of engagement, protest, and action at length in the coming sections, along with different modes of participation.

Within participatory culture, the structures of the architectures of participation have five forms, according to Casey Reas. They are repetition, transformation, parameterization, visualization, and simulation. Repetition and transformation are integral to the process of remixing itself, while the other three are required for the creation of mediated disturbances, which include identifying problems, devising strategies of action, pattern recognition, and the ability to make sense of large quantities of data. These last three are increasingly becoming criminalized, as arrests of Julian Assange of WikiLeaks and Aaron Schwartz at MIT (arrested for downloading too many articles from JSTOR, the academic database) attest. Before I discuss these tactical aesthetics at length, first a discussion of the creative act itself is necessary.

Long live the reflexive remix

Celebrated French filmmaker Chris Marker is one of the most inventive and creative minds of our time. To mark the centenary of film, Marker produced a work called *Silent Movie* (2005). It comprises five stacked televisions – with the center one, "Captions," playing 94 silent movie intertitles from unknown or imaginary films. The other four televisions play videos simultaneously with four visual themes: "The Journey," "The Face," The Gesture," and "The Waltz." The 'data' or content are drawn randomly from five laser disc players, which hold 20 minutes of video each, shot in the style of silent film, and a computer interface connects it all together and randomizes the clips. The installation is complemented by 18 black and white video stills, 10 film posters and a silent movie soundtrack of solo piano pieces that play throughout for a total of 59 minutes and 32 seconds. Marker's work is a simulacra emulating the aesthetic of silent film with his own shot footage of actress Catherine Belkhodja. By splitting it up on to multiple screens or 'channels' he spatializes the work and makes apparent the fragmentation that was in montage-rich, narrative film all along. The actual remixed content is in the music and posters that surround the installation of televisions. The soundtrack, "The Perfect Tapeur," is a medly of solo piano pieces drawn from the compositions of silent film composers, "Bill Evans, Alexander Scriabin, Billy Strayhorn and Nino Rota" (Seid, 1995). The music lends emotional fixity to the jumble of random images that are generated by the computer program. The posters too are a simulacra of sorts. They are posters in the style of silent films for films that might have (*should have*) existed, including:

A silent version of *Hiroshima, Mon Amour* starring Greta Garbo and Sessue Hayakawa; Ernst Lubitsch's *Remembrance of Things Past* starring Gloria Swanson, John Barrymore and Roman Navarro ("The first movie where the captions take more space than the image"); and Oliver Stone, Sr.'s *It's a Mad Mad Mad Dog* make up just part of Marker's alternative cinema history (Sied, 1995).

Marker describes the work as the "pre-historic state of film memory" (Sied, 1995). This is what film remembers of its old life as a silent form.

Despite all of the remixed and derivative material, it is easier to see the originality of Marker's resituation, reinterpretation and revisioning of a bygone form in *Silent Movie* than it is in Wilhelm Susnal's *Untitled (Elvis)* (2003), but there is common ground in their works. Susnal's *Untitled (Elvis)* is in three parts. The video begins with a collage of Elvis clips playing on a computer. The computer itself is spinning on a turntable around a dangling microphone, in front of a window. Not only is this montage a series of stitched-together jump cuts of "Tell Me Why" from different concerts in Elvis's early career, but our act of seeing it is constantly interrupted as the turntable spins. The second part shows Elvis at his last concert singing *Unchained Melody* (1977). Susnal keeps the online browser interface in the frame with identifying account names – it might be a YouTube URL – and other information 'censored' with thick black lines. The camera slowly zooms in on the image and we watch the remainder of the song without commentary, but with the browser window still visible. The third part of the film cuts to filmed footage of another Web posting, this one of an older Daniel Johnston, the Austin outsider artist-singer who has spent so much of his life struggling with bipolar-related issues, performing his "Caspar The Friendly Ghost" song. The first two parts are connected by the continuity of Elvis material. The third part is only tangentially linked by the theme of fan footage of a sweaty, aging, overweight singer. Like Elvis at his last concert, Johnston is past his prime and trapped within the perils and pitfalls of a celebrity culture that consumes the most talented. With Susnal's film as a critique on continuity and performance, we look and look again. Like Marker's work, it is multichanneled and requires our interpretation in space to make sense of the modular parts.

The difference between the Disneys on the one hand and the Markers' and the Susnals' remix techniques on the other is that, unlike artists, Hollywood absolutely plagiarizes. It plagiarizes itself. The machinery of Tinseltown produces versions of the same film over and over and over again. Celebrating the genre, pastiche, and imitative style as creative techniques, the large Hollywood studios cultivate a viewing public that is hungry for particular types of films or franchises. Remakes and sequels and tried-and-tested formulaic recipes abound. In his article "Innovation and Repetition: Between Modern and Post-Modern Aesthetics," Umberto Eco lays out a formula for explaining the formulaic nature of Hollywood film where, he says, "iteration and repetition seem to dominate the whole world of artistic creativity" (Eco, 1985, 166). For Eco, mass media were built on the devices of "repetition, iteration, obedience to a pre-established schema, and redundancy (as opposed to information)" (1985, 162). The device of iteration is a kind of formula that guarantees the delivery of a punch line, as in Sherlock Holmes's trademark unraveling and explication of a mystery in the final scene. Under postmodern aesthetics, Eco's modes of imitative creation include the retake, the remake, and the series, and among different kinds of series are the flashback, the loop, the spiral (strips), the saga, and intertextual dialogue. The series, for instance, revolves around a cluster of fixed characters and a schema for the story that is predictable, like *The Terminator* series. The flashback is a little different; it creates a space for character development rather than plot advancement as in the ever more complicated series *Lost*. The spiral is a model most often used in comic strips where nothing ever changes and characters never age or grow up. The most successful mass media forms establish a dialectic between new and old material and set out an interplay between form and content with which the reader/viewer is already familiar. Once these two conditions are met, the greater the aesthetic value of the work. The sophisticated reader or viewer of such works develops a sort of double vision, Eco maintains, reading for plot on one level and for aesthetic merit on another; the former enjoys the formula and the latter the differences.

Hollywood does not just copy itself either; there occurs a rich global cross-pollination between East and West and North and South. Think, for instance, of Japanese master director Akira Kurosawa's *Yojimbo* (*The Bodyguard*; 1961); it sets silk merchants versus saki merchants and was so popular it spawned two Japanese sequels, *Sanjuro* (1962) and *Zataichi Meets Yojimbo* (1970). But the Japanese were not the only ones watching. Italian director Leone transposed *Yojimbo* to the American west as an English-language western where he set gun merchants versus liquor merchants and called it *Fistful of Dollars* (1964). It was filmed in Italy and starred a young, unknown actor, Clint Eastwood. That film was so successful that it created a whole new genre, the spaghetti western, and two new sequels also with Eastwood, *For a Few Dollars More* (1965) and *The Good, The Bad and The Ugly* (1966). The spaghetti western was so popular that Leone's work was widely emulated in more than 30 Italian copycat films, the most famous of which was *Django* (Corbucci, 1966). *Django* sets Mexican bandits versus sadistic Gringo vigilantes. The film was so similar to the other films that its release was delayed for a year in the U.S.A. due to copyright infringement issues. Many other films were inspired by that film, including Australian Road Warrior series, which starred a young unknown actor, Mel Gibson, as *Mad Max* (Miller, 1979, 1981; Miller and Ogilvie, 1985), which set drillers versus bikers. Other remakes followed including the gangster flick *Last Man Standing* (Walter Hill, 1996), which featured Bruce Willis, bootlegging Italians, and the Irish feuding in Texas. *Lucky Number Slevin* (Paul McGuigan, 2006) set the Black mafia against the Jewish mafia. Then the story came full circle and Japanese director Takashi Miike filmed his pastiche masterpiece *Sukiyaki Western Django* (2007). Miike's rendition sets the Heikes (the Reds) versus the Genjis (the Whites) in an asianfied retelling of the War of the Roses. Then, just when we thought we'd seen it all, Gore Verbinski of *Pirates of the Caribbean* franchise directed *Rango* (2011), an animated bird, animal, and reptile-populated western with Johnny Depp as a chameleon loner. But that is not the end of the story. For all that Kurosawa is credited with crafting a popular tale that was so successful it was endlessly emulated, it is important to note that Kurosawa's original *Yojimbo* script was an adaptation of American novelist Dashiell Hammett's *Red Harvest*. This is not a unique phenomenon. The quintessential American urban post-war genre *film noir* was a remix of German expressionist film techniques as realized by ex-patriot German, Scandinavian, and British directors. The monster genre of disaster films owes its origins to the Japanese 21-film franchise *Gojira* (Honda Inoshiro, 1954) (released in the West as *Godzilla, King of the Monsters* (1956) with the substitution of Canadian actor Raymond Burr for some of the Japanese actors and footage).[2] The whole Kung Fu genre is the result of Hollywood emulations of Chinese and Japanese films. Kung Fu films spawned dozens of blockbusters, many with Asian actors in lead roles, like Jackie Chan and Chow-Yun Fat, or directors at the helm, like John Woo. Where *auteur* Quentin Tarrantino can remix the Japanese samurai genre to produce *Kill Bill, Parts I and II* (2003 and 2004), Taiwanese-American director Ang Lee can reinterpret the western in his 'all-American' academy award-winning gay western called *Brokeback Mountain* (2005). Manga and anime have spawned a whole new global industry, including left-to-right reading comics in the West. In addition, the influence of Chinese films' elaborate wire-work, temporal shifts, and time-enhancing special effects have given birth to a new kind of high-tech science-fiction-kung-fu mashup genre, which includes *The Matrix* series (Wachowski, 1999). That in turn paved the way for Chinese films made not just for a home market but with an eye to international markets to become major international hits including Ang Lee's earlier film *Crouching Tiger, Hidden Dragon* (2000), and Zhang Yimou's *Hero* (2002) and *House of Flying Daggers* (2004).[3]

What is happening with new media artists appears on the surface to be similar with repetitive iterations everywhere. But surfaces can be deceiving.

The reiterated material is only the content and it is a vehicle for critique, not the *raison d'être* for the work itself. Remix is the "transgressive and critical manipulation of media technologies" (McLeod and Kuenzli, 2011, 13). Remixes often critique the media themselves as well as adapting content to explore meanings buried beneath the surface. Where Walt Disney unapologetically took Keaton's steamboat captain (presumably as an act of homage) to create Steamboat Willy, in our time artists reuse other works to reveal or transmit political messages, and not simply to re-create them. That retelling is critical, deeper, and multilayered, asking us to rethink the original through the new. In stark contrast to Steamboat Willy, for instance, British filmmaker Steve McQueen produced a video installation piece that is a re-enactment of Keaton's most famous scene, a scene where the façade of a house falls and leaves Keaton unharmed and upright in the space where the window had been. McQueen's *Deadpan* (1997, silent, black and white film transferred to video) is not a re-creation of the whole of Keaton's *Steamboat Bill, Jr.* (1928) or even the theft of a character, as in Disney's case. It is the recontextualization of that scene without narrative and in a different genre. The original was a long shot of a visual gag, but McQueen's new version splinters the perspective. It is shot twice from two different perspectives – the first as a medium shot taken from the back and the second as a close-up of McQueen's boots and legs shot from the front. It is a perfectly looped version of the house front falling, with the terrible gravity of the situation reinserted. We never see McQueen's face in the video. We only see his tensed back in the first scene. In the second scene, we see McQueen's feet and his flinch as the house falls. McQueen's and our awareness of his own mortality and his fear become players in the video. Another layer of meaning is inserted in the remake by the change of actor or, more specifically, by the change of the actor's race. Keaton was small in stature, skinny, wiry, and pasty white in complexion. McQueen is a large, broad-shouldered, well-built black man. McQueen's white shirt and jeans and his facelessness in the film transform him into what seems at first glance to be a laborer, into a nameless, faceless construction worker with all of the politics and assumptions of race that are implied by that. Initially then, we might be inclined to read the looping scene as a construction accident. That is, we might read it that way until we see the second clip with its close-up of his laceless shoes. This raises an even greater specter of racialized type: is the black man an (escaped) convict? We know from the aesthetics of silent film that this is a chase, since the looping chase was an integral visual component of the silent era. Immediately, then, the laceless boots suggest someone on the run, even though, unlike Keaton's original, McQueen makes no movement to run away after the accident.

Ursula Frohne in "Dissolution of the Frame: Immersion and Participation in Video Installations" locates the significance of McQueen's film in the temporal and perspectival rupture. Such a remake results in the splitting of our point of view: "The viewer is caught somewhere between the past memory of the original silent film, the phenomenological time of the sequence's reconstruction and the time and space of the viewer's own physical reality" (quoted in Frohne, 2008, 364). Furthermore, it is in the act of imitation, of 'seeing again' that reveals the illusion and the differences in the original. The imitation, she says, reveals the otherwise invisible boundaries between art and life, between the artifice of lifelike reality and the screen (2008, 364). Marker, McQueen, and Susnal's works are not *digital* video, but all filmed on film and then manipulated digitally – by transfer in McQueen's case, as a subject in Susnal's, and as a means of playback, randomizing, assembling, and structuring by Marker. None of them are untouched by the digital. Everything is now digital, and that leads to strange cross-pollenizations and hybridizations on screen, in the gallery, and in public space. According to Bruce Sterling, we live in a world of gothic absences and leftover high-tech realities where, he says, there is no channel, no film, no medium. While this is true on some level, the technologies and their specificities also persist as ghosts in the machine and as modes of distribution and display. Their aesthetic attributes linger on. Marker, Susnal, and McQueen's works can come about as a result of the meeting of analog video in a computerized age, a medium and an age that share certain similar qualities with digital technologies. Analog video was revolutionary because it used live capture. Unlike film, no processing is involved to view it again either as it is recording or after the fact. The magnitude of this shift from film to video in terms of the nature of the art form cannot be underestimated.[4] In addition, video was what McLuhan would call a 'cool' medium – a medium that was low in information, stimulated multiple senses, required sensory involvement, and so we had to insert ourselves into it to interpret it, like comics. As forms of interruption, Marker (with his computerized randomizer), Susnal (with his browser windows), and McQueen (with his continuous, stuttering loop) evidence the aesthetics that will become so prominent in digital technologies. These works also demonstrate qualities of interaction that we do not see in cinematic film and make plain that many of the techniques digital practitioners are using in new media are not new at all. As well, there is a slippage. Once *on* the screen or monitor, the origins of film, video, and digital video can become indistinguishable to the untrained eye. What is new in the digital universe is the ability to work with exact copies, with precisely the same image. The photograph, the Xerox, and the cassette tape started this process of capture, but the simulacra of digital – where everything is a copy – has changed the stakes forever.

In an age where almost anyone can have access to powerful digital technologies, our media engagement is entirely altered. The convergence of the acts of watching, producing, and distributing in one machine has given everyone the potential to be an author (and as technologies get cheaper, smaller, wireless, and more mobile this trend continues to accelerate). When everyone can be an author, where does artistry reside? What is authorship if it is no longer an exclusive club? Where lies authority? The best 'new' (remixed) work is not merely transformative, but operates within a conceptual framework. Consider, for instance, *Violent Paintings* by the Italian, American, and German digital collective Alteraziona Video (2009/10).[5] Working with images found on the Web, these so-called paintings are found digital images, manipulated many times by each member of the collective, to be ultimately printed on warped JVC and displayed in galleries. "*Violent Paintings* are aesthetic scrap created by altering images found on-line on amateur sites, a hotchpotch of home photography, pornography, *hentai* graphics, second-rate special effects and art classics" (*Piemonte Share*): "The group acts as a widespread, mobile international network exploring misinformation and the relationship between truth and representation, legality and illegality, and freedom and censorship, combining the languages of art with political activism across all forms of media" (*Piemonte Share*). The damage that is done to these images (in the presence of the images' banal nature, the notion of 'violence' is clearly tongue-in-cheek) via editing software, photoshopping, media transfer, printing, and translation cannot mask their lowbrow and hackneyed nature, but "the actual violence, as the artists write is what is inflicted on the spectator 'who is challenged to make some sense of the big jam of amateurish and pornographic images, of cheesy special effects and reproductions of artworks'" (Caronia, 2010). This form of digital manipulation was not born from popular forms, but digitally has its roots in conceptual art.

Conceptual art is often political and situated, and privileges concepts or ideas over the aesthetic and material as a way of questioning the nature of art itself. As Walter Benjamin predicted, in the age of post-mechanical reproduction the work of art becomes "designed for reproducibility" rather than for a sense of uniqueness (Benjamin, 1968, 224). In the 1960s the philosophy that "the idea is the work of art" (Alberro, 2003, 160) and "everyone's an artist" (Alberro, 2003, 163) arose out of New York curator and dealer Seth Siegelaub's conceptual gallery shows. By reframing art outside of the sphere of representation, conceptualism and its descendants seek to break with institutions, social structures, and identity constructs all in one go. As the public sphere has grown increasingly narrow with the rise of consumerism and the corporatization of copyright, conceptual modes of discourse offer possibilities for new collaborative practices and new community formations.

The conceptual art frame or aspect in the collaboratively created *Violent Paintings* is in the rendering of content as data. In our time, the image's content has, in short, become irrelevant as the object has dematerialized. The subject itself of remixed works is the medium of its end-product and the self-reflexive way it draws the user in to interact with it. In "Shadows of Video Art: The New Era of Quite (*sic*) Images,"[6] Caronia argues that the overwhelming deluge of information that is now available to us renders the need to produce new images redundant, superfluous, and irrelevant. Instead the new goal for artists is in processing. The new triumvirate of digital creation, Caronia says, is "reading, editing and distortion processing," for, "When everybody is producing media, selection becomes the only way of producing content" (Quaranta, 2011). Caronia goes on to say, "The very moment when the availability of technologies of production and publication soars the number of image creators, the figure of the author staggers. When everyone produces media, hence becoming a media 'author,' there is no author anymore" (Caronia, 2010, n.p.). The concept of the author remains intact only in the art world where big bucks ride on the brand of a name, but that is just about the only place it is (and, according to Alexander Alberro in *Conceptual Art and the Politics of Publicity*, it does so at the expense of conceptual art). Even in the world of high art the concept of authorship may be somewhat shaky as stardom, celebrity, and outrageously inflated price tags are redefining and threatening to replace authorship in media culture right across the board.[7]

Alterazoni Video is not the only group that is undertaking such experiments either. Christine Paul links digital arts' antecedents to Dada, Fluxus, and conceptual art. She says, "The importance of these movements for digital art resides in their emphasis on formal instructions and in their focus on concept, event, and audience participation, as opposed to unified material objects" (Paul, 2008, 11). The collective Jodi, an Internet pioneering duo made up of Dutch artist/programmer Joan Heemskerk and Belgian artist/programmer Dirk Paesmans, has been creating net.art along these lines using screen grabs, code errors, viruses, game mods, and other detritus as raw material for digital art since the mid-1990s. California artist Petra Cortright uses video to question spectatorship and interrupts the viewing to make the viewer aware of the monitor. The Mexican-born New York City-based artist Brody Condon resituates images of video game trauma in installations, video, and performance as in *Death Animations, 2007–2008. Death Animations* is a giant collaged riff on Bruce Nauman's "Tony Sinking into the Floor, Face Up and Face Down" with Japanese and barbarian game characters seen "through the lens of 'New Age' ideas such as astral projection … and recent foreign conflict" (Condon, 2011). Condon's *KarmaPhysics < Elvis* (2004) uses a game mod of a particularly gory science fiction game *Unreal*:

When plugged into a projector/monitor and power, a small custom pink computer automatically starts and displays the work. The viewer is pulled slowly through an infinite pink fog filled with floating, twitching bodies of Elvis Presley. The convulsions of Elvis are controlled by the original game's Karma Ragdoll real-time physics system – generally used to simulate the physical dynamics of game character death (Condon, 2004).

Elvis's characteristically twitchy arms, legs, and pelvis become something quite disturbing when resituated in this context. Another video artist, Austrian Oliver Laric, does fascinating work exploring echoes of images in other images and on the derivative nature of our culture.

In Laric's video *Versions 2010,* he parallels scenes from *Winnie the Pooh* and *The Jungle Book*, for instance, where whole sections of the later film have been taken cell by cell from the former and redrawn with different scenery and color modifications; he plays side by side scenes from films that use similar shots in the same locations; he shows us multiple versions of sculptures from antiquity. In Laric's video, the voiceover (which may or may not be culled from other sources) states,

> [T]here is more work in interpreting interpretations than in interpreting things and more books about books than any other subject. Multiplication of an icon, far from diluting its power, rather increases its fame, and each image, however imperfect, can eventually pass conventionally partakes of some properties of the precursor. [...] The more images, mediations, inter-mediaries, icons are multiplied and overtly fabricated, explicitly and publicly constructed, the more respect I have for their capacities to welcome, to gather, to recollect meaning and sanctity (Laric, 2011).

Some of his other exhibitions include:

> "The Real Thing" after a short story by Henry James on an artist who prefers representation over reality. We showed a series of press releases selected by Daniel Baumann, tourist photographs of Jeff Koons's *Balloon Dog* at Versailles, a *New York Times* back issue with an article debating the existence of Gelitin's balcony mounted on the World Trade Center in 2000, a video sampler of Seth Price's editioned videos, a PDF version of a performative talk by Cory Arcangel, acoustic versions of Claude Closky's text pieces, and cam versions of the current Hollywood blockbusters, among other works (Quaranta, 2011).

One of the most lucid commentators on the nature of authorship in the history of art and popular culture, Loric uses video art to interrogate the act of creation and the art of inspiration. For him, no source material is off limits.

It is clear that for Loric and these other artists a large part of their creative process is now spent *locating material* to adapt, rework, mashup, and alter. Where these artists critique and explore the creative act and spectatorship, other artists have undertaken much more radical experiments in terms of reinventing authorship. Andruid Kerne with his *CollageMachine* (1998) and Judd Morrissey and Mark Jeffrey's *theprecession.org* (2011), for example, seek to narrow the divide between the audience and the author. Others, like the duo Soda_Jerk in *Pixel Pirate 2: Attack of the Astro Elvis Video Clone* (an "anti-copyright epic"), and Pierre Hughye with Philippe Parreno in the manga *No Ghost Just A Shell* (2000) have critiqued authorship within the framework of copyright itself. I would next like to take a deeper look at the three modes of creative critique: interruption, disturbance, and capture/leakage.

Interruption (stoppage + repetition)

As lawmakers try increasingly to criminalize individual creation, open access, and downloading, it is clear that they cannot stop the flood of images, works, and ideas that are being remixed. What is misunderstood (perhaps willfully) by corporations and lawyers is the fact that the image or text itself is no longer either the subject of the work or the art object itself. Product has been replaced by process. Recycled images or fragments of text are literally hiccups in the datastream designed to catch our attention. It is the datastream in conjunction with documentation that is the 'new' original work. In a time when remixing, sampling, and mashups have become the definitive creative acts, the seminal component on which the critique hangs is the interruption or stoppage. It is in the rupture, pause, and repetition that works of art are now being born. 'Stoppage' and 'repetition' – the terminology of Italian political philosopher Giorgio Agamben – are the spark that give birth to critical awareness. It is in that hesitation between image and meaning that the power of the remix is unleashed. It is in the hinge between stoppage and repetition that a door is opened for the reader or spectator to insert herself and become an active participant (Agamben, 2008, 332).[1] The gap, the pause, the space of critical thought for poet Paul Valéry also existed in the spaces of poetic language. He defined the poem as "a prolonged hestitation between sound and meaning" (quoted in Agamben, 2008).[2]

Agamben believes it is in this stop that the word or mode of expression makes its own representation visible. This is especially important in our age of post-representation, emergent as a result of the politics of difference, where it is no longer acceptable to be judged by our skin color or gender or sexual preference. Post-representation foregrounds *différence* as difference that is visible but not a disqualifying feature. That is to say, difference is acknowledged and embraced while humanity is affirmed. Similar to Eco's double vision, Sonia Kruks explains:

> What makes identity politics a significant departure from earlier, pre-identarian forms of the politics of recognition is its demand for recognition on the basis of the very grounds on which recognition has previously been denied: it is *qua* women, *qua* blacks, *qua* lesbians that groups demand recognition. The demand is not for inclusion within the fold of "universal humankind" on the basis of shared human attributes; nor is it for respect "in spite of" one's differences. Rather, what is demanded is respect for oneself *as* different (2001, 85).

Post-representation therefore moves away from categorization or identification on the basis of the visible image or a surface, but also affirms the image as a defining attribute. So it both is and is not. This is similar to simulation – the creation of a world in a virtual environment that copies the real but has no real life referent – but with a significant difference. A representation can be a simulation, but not all simulations are representations. William T. Mitchell makes a distinction between the cinematic image as *representational* and the simulated digital image as *presentational* (Paul, 2008, 86), the latter being more like scientific data.

The repetition of the familiar in art has gone by many names. When German artist Kurt Schwitters called his architectural, collage-based, art practice *Merz*, and French theorist Nicolas Bourriaud refers to remixing as *postproduction*, both referred to the practice of reusing, quoting, and resituating pre-existing materials. Similarly, repetition, Agamben and many others argue, is not the same thing seen a second time. Repetition is the rediscovery of the familiar in the new. It is a seeing again. It is the return of the idea of something like a memory made tangible. Agamben was speaking in reference to French theorist, filmmaker, and revolutionary Guy Debord's moving images specifically, and it was the use of montage in Debord's works that most fascinated Agamben.

In Debord's cinema, Agamben finds Valéry's philosophy manifest; he defines it as a "certain kind of cinema [that] is a prolonged hesitation between image and meaning" (Agamben, 2008, 332). It is easier still to see how this hesitation applies to digital media that celebrates rupture with its fragmented lurch and jerk between disconnected, hyperlinked, or parallel screens.[3] Sometimes the break is also between different versions of a work, as in Finnish artist Eija-Lissa Ahtila's representation of the works in different media and settings. She varies the use of single and multiple screens (sometimes in immersive installations, sometimes in cinemas) for vastly different effects.[4] We also see this disjuncture between literal screens in gallery space as with Marker's five-monitor installation, Chinese artist and gamemaker Feng Mengbo's parallel screens in *The Long March,* Iranian video artist Sherin Neshat's complex split-vision filmic installations where genders are segregated on facing walls, and in German artist Harun Farocki's[5] multi-screen installations. According to Agamben, the expressive act itself is fulfilled when the medium disappears, when we see only the contents of a medium and no container (2008, 333). Conversely, stoppage and repetition's unique attributes serve to make the medium visible. The pause and replay are break boundaries that make the image disappear so that the medium itself becomes apparent to us. A break boundary is something that makes a medium visible – for a medium is invisible (like water is to fish) – and that phenomena can suddenly become apparent, like sound waves on a plane's wing. Guy Debord, who used stoppage and repetition as exemplary methods for performing his ideas, was a master artivist and theorist. It was in just such a gap, between the absence of representation and the withering of the art object, that conceptual art protested the rising consumerism and made the environment or political structure of the gallery visible in the late 1960s.

Guy Debord was the most prominent of the Situationist International (1957–72) philosopher-artivists. Debord was a revolutionary who used constructed situations in time through rhetorical *dérive* and the *détourning* of images and texts, alongside psychogeographic explorations of urban space to critique the anaestheticization of the senses that he saw and feared in the arrival of mass media and visual culture. In his influential Marxist treatis, *Society of the Spectacle* (1967), the book and the later film of the same name (1973), Debord explores the vampiric and anaesthetizing effects of mediated consumer culture. The "spectacle" is a visual means of social domination that wrought the transition from the labor of the industrial age to the consumer culture age's utopian dreams of fulfillment through consumption. He and the other Situationists articulated an anti-aesthetic (or an anaesthetic?) that called for the devaluation or even suppression of highbrow art in order to assign art a whole new role. In its place, a tactical craft would replace it – what they called "a new genre of creation" (*Détournement as Negation and Prelude,* 1959). Their highest goal was that creative practice and culture be integrated into everyday life. Consumerism had won the war, as they saw it, of turning active subjects into passive consumers through a drowning of real experience in mediation, and through the promotion of inauthentic trappings of lifestyles over authentic experience. If refusal and resistance were to be expressed, it must be done spontaneously, Debord argued, in order to create nonrepresentational *situations*. Situations were a process or the antithesis of object-based art. The rise of the spectacle was contemporaneous with the arrival of consumers. Debord saw in the shift from Karl Marx's warnings of the implications of a commodity culture to an inevitable commodification of humanity ready to be auctioned off to the highest bidder through "a constant deluge of mesmerizing images (produced by advertising, Hollywood, television)" in lieu of real human exchange (Harold, 2007, 3). This new concern with surfaces and superficiality was a result of the redefinition of the role of the human subject: as someone who *had* rather than *was*, and who opted for lifestyle choices over ownership. The "degradation of being into having" and ultimately by a "generalized sliding of having into appearing" (Harold, 2007, 3) was insidious and impossible to resist from within the Spectacle.

Debord and other Situationists are credited with having played a key role in the May 1968 uprisings and general strike in France, which united 10 million workers, students, and intellectuals, and overthrew the French government. In her classic work *May '68 and Film* (1980), Sylvia Harvey credits these events with being "an 'impetus' to the burgeoning discussions of ideology in general" and, along with Althusser's constructed "subject" (who is a product of ideological state apparatuses), led to discussions of spectatorship, and "the potential of radical filmmaking in particular" (Aaron, 2007, 5). Debord as an experimental filmmaker was a key player here. Rebelling against a political framework that rooted all relationships in 'transactions,' Situationism as applied philosophy set out to create situations outside of representation in visual culture in which "a moment of life, concretely and deliberately constructed by the collective organization of a unitary environment and a game or the free play of events" (Debord, 1958). The methods by which a "situation," "environment," and a game or "events" were constructed were threefold. Situationists' tools were *dérive*, *détournement*, and the creative use of urban space and architectural reform through psychogeographic wanderings. Intensely aware of the role of the visual, they sought to "provoke people" into a "rediscovery of joy and a reversal of perspective" (Harold 7). They were the original hackers. They were a group who yearned for the breakdown of hierarchies in order to craft a networked society with the fluid organization of peer-to-peer communities. As Debord says in Thesis 24 of his book, "the society of the spectacle is a form that chooses its own technical content." This content was what we would call remixed material. In 1956, Guy Debord prophetically described the remix process: "The first consequences will be the revival of old, cheap, bad materials and extensive (unintended) participation of their unknown authors" ("Methods of *Détournement*"). In their advocation for a dialogue in lowbrow art made by everyone, we can see the seeds of social media. In favor of vandalism, sabotage, and full-scale revolution, the Situationists sought to lose themselves in the drift of *dérive* and to seek to reverse, hijack, and redirect political (specifically capitalist) systems through *détournement*. "*Détournement* was an effort to devalue the currency of the Spectacle" (Harold, 2007, 7) and to replace it with a creative process. They disavowed the creative act itself and previous 'inspired' works (what we might call canonical art and literature) as "obstacles, dangerous habits"; they believed "plagiarism is necessary, progress implies it" ("Methods of *Détournement*"). Instead they advocated remixing lowbrow, existing materials. In their post-creative world, they seemed to foresee what we would now call remixes and mashups, since, through its doubling of meanings, *détournement* diverts and misappropriates to foster a new genre of creative output. In "Methods of *Détournement*," the Situationists described this new method of creating art this way:

> Any elements, no matter where they are taken from, can serve in making new combinations. The discoveries of modern poetry, regarding the analogical structure of images demonstrate that when two objects are brought together, no matter how far apart their original contexts may be, a relationship is always formed. Restricting oneself to a personal arrangement of words is mere convention. The mutual interference of two worlds of feeling, or the bringing together of two independent expressions, supercedes the original element and produces a synthetic organization of greater efficacy. Anything can be used (Debord, 1956).

This is the interruption where stoppage meets repetition to form new meanings, and in that twist there are three fundamental algorithms to *détournement*: (1) It is a literary/artistic communism that erases hierarchies in images because no single element has greater importance than another. The goal is to make the manipulation apparent, like in culture jamming. (2) It creates a new, fresh meaningful assemblage. I believe that *détournement* is at its roots a method to enable a social mobilization. It is not designed to foment revolution (which would simply reverse existing hierarchies) but to cause a disturbance instead. (3) *Détournement* is activism. It is social, contextual, and situated. It is a tactic geared to a particular, like-minded audience ("Methods of *détournement*") and it requires participatory activity. It is a dynamic environment or information space – a pressure cooker for ideas. Their methodologies of *détournement* were remix techniques in four parts. First, to create dissonance through jump cuts, leaps of logic, and the link to incongruous things. This is what we would now call a mashup, like

Danger Mouse's *Grey Album,* which is a blending of the Beatles' *White Album* and Jay-Z's *Black Album.* This was an inspired creative venture that by some accounts was the most successful American album of 2005,[6] even though it was never sold (for legal reasons). Second, simplicity springs from reusing familiar materials, and it is the recollection of the original contexts in their elements underlined by changes that stops us in our tracks (this is the process of interruption in the stoppage and repetition that I have been discussing[7]). A prime example would be Antonio Mendoza's "State of the Union" (2006). Taking one of George W. Bush's State of the Union addresses as a starting point, Mendoza has selected all of the pauses and moments when Bush is about to speak. The original text is then interrupted and repeated, so the address becomes a series of rapid-fire stutters, twitches, pauses, blinks, and fragmentations like a scratched record or a beat-juggled piece of music. Eventually even the picture itself breaks up into giant rhythmic accordion pixilated patterns, representing a metaphorical media implosion or the complete breakdown of communication. A more highbrow rendition of this technique may be seen in David Tevita Siufanga's found footage installations. For instance, in one untitled piece (2010; http://vimeo.com/14991008) he juxtaposes a bank of four monitors, each playing a single brief loop of filmed action; from right to left they are from *The Last Samurai* (Zwick, 2003; about an American who embraces the Samurai culture he was hired to destroy), *Blood In, Blood Out* (Hackford, 1993; based on the true life experiences of former LA Vatos Locos gang member and poet Jimmy Santiago Baca), *The Godfather* (Coppola, 1972), and a home video shot at the zoo found on YouTube. The first screen slowly zooms in and out on a lone samurai standing in a wheat field; the second shows gang members grabbing weapons at the start of a fight; the third shows a loop of a hated male character stepping in and reversing out of a doorway; the fourth shows a lion roaring. Together the four video loops, which play backwards and forwards repeatedly for a total of 2 minutes and 39 seconds, create a rhythmic symphonic response to menace and danger meeting in a cacophony of stillness and flight, fear and aggression, stoicism and dignity. The third methodology is that these forms themselves are illogical, irrational, and irreducible to a linear argument. One of the signature graffiti-scrawled slogans from the May 1968 uprising, for instance, was "Do not adjust your mind, there is a fault with reality." (Bray; Here we can see the genealogy of these ideas too as Jean Baudrillard built on the Situationists' thinking and the Wachowski Brothers adopted Baudrillard's ideas for their films in *The Matrix* series.) The final methodology is the avoidance of simple reversal, which is least effective. Instead there is a need to reroute logic in surprising ways like McLuhan's probes – a recognition of the fact that simple negation is a dead-end. Ad busting and culture jamming fall victim to this binary form of argumentation and therefore merely comment on rather than enact a revolution of everyday life. Contemporary practices of billboard modification are a little freer and can either reimagine or merely reinscribe existing codes.

The Situationists, like the Frankfurt School before them,[8] "believed that the spectacle had become so sophisticated that it could successfully recuperate rebellion, strip it of its threatening content, and resell it as pure image" (quoted in Harold, 2007, 12) – just like elitist pop artists did after them. Time has proved them right. Situationists were not fans of the remixing of existing images, although Debord did just that in his films. In fact, they were not fans of photographic (i.e., mechanically produced) images at all. *Détournement* does not simply "negate the negation" equation (Harold 13); it inverts it and always adds a spin or a twist. In her book *OurSpace: Resisting the Corporate Control of Culture,* Christine Harold says this is a weakness in the Situationists' philosophy. She says that always negating the negation, one can never *détourne* sufficiently (Harold, 2007, 13), but that is only the case if one is looking for *products* as an end result rather than at the *process*, as a continuous action. *Détournement* is at its best a never-ending conversation, as in the work of Italian artist Maurizio Cattelan. In popular culture, there is also dialectical activity around familiar works. Look, for instance, at American pop star Beyoncé's 2009 music video, "Single Ladies (Put A Ring On It)." One of the most popular YouTube videos, it has been viewed more than 132 million times over a three-year period. It is also a video that is highly emulated, imitated, and parodied. People post videos of their children dancing to it, or demonstrate their dancing prowess in their own versions. Another viral video, that of Radiohead's lead singer Thom Yorke's "Lotus Flower," has, among many other Beyoncé-inspired versions, also been recut to the "Single Ladies" soundtrack. One parodic mashup with three drag queens by Rederecord has received 7.2 million hits of its own; that version is itself a remix of the *Saturday Night Live* version of the performance where Beyoncé danced with a high-heeled, leotarded Justin Timberlake, Joe Jonas, and another dude. Even the dishwater-dull "Double Dream Hands" videos, by Beyoncé's choreographer John Jacobson, that demonstrate the moves step by step, have received millions of hits, and the Jacobson video itself continues to be remixed (like a mashup Bollywood version) to boot. Beyoncé's original song is a cry of defiance for a single woman asserting herself and her own self-worth after being rejected by a lover. For all its feminist and feel-good content though, ultimately in the song she is still willing to offer herself up for ownership in exchange for a ring. Award-winning Canadian cultural theorist, poet, performance artist, and New York City Professor of Global Literature Adeena Karasick uses Beyoncé's video as the framework for a rallying cry for women to create conceptual poetry, viral videos, and other subversive literary and artistic creations.

The lyrics for "Single Ladies (Put A Ring On It)" are a battle cry of sorts. Beyoncé invokes all the single ladies to identify themselves on the dance floor:

> Now put your hands up
> Up in the club (club)
> Just broke up (up)
> Doing my own little thing
> You decided to dip (dip)
> And now you wanna trip (trip)
> Cuz another brother noticed me
> I'm up on him (him), he up on me (me)
> Don't pay him any attention
> Done cried my tears (tears), for three good years (years)
> Ya can't be mad at me.

The lyrics thereby are a statement of absolution, giving herself permission to start to date again, but always in the song is the awareness of the old boyfriend's eyes being on her and of her needing to explain or justify her actions. The refrain in the original song is "'Cuz if you liked it then you shoulda put a ring on it/.../Don't be mad once you see that he want it/If you liked it then you shoulda put a ring on it." Karasick takes this foundation and *détournes* it with her YouTube remix. In her hands, this becomes a rallying call for conceptual poets to get down to the business of reseeing, of rewriting, of blocking and subverting original meanings. She calls on "All the lingual ladies" to:

> Put your pens up
> Up in the hub (hub)
> Where the memes break up (up)
> I'm doing my own little thing
> Decided to rip (rip)
> The ambiguous script (script)
> A stutter flutter of floating memes
> I'm up on it (it), it up on me (me)
> Just give it all of your attention
> Don't be full of fear (fear), just frame what you hear (hear)
> Like a text of radical memes (memes).

A meme is a cultural contagion. By definition, they are small, sticky, self-replicating or mutating concepts or ideas that circulate via a medium (gesture, ritual, speech, books, the Web). According to Susan Blackmore, there are two different ways in which memes transmit themselves. They can either replicate themselves as a product or they can copy the instructions of their own creation (Blackmore, 1999, 288). The latter is an algorithm, a dynamic line of code, that would have appealed to the Situationists very much and it is the latter that Karasick is enacting in her *détourne* of "Single Ladies." The speaker thereby unleashes her radical and subversive ideas by stepping outside of consumer culture as an act of political critique through her viral text. Following the refrain "'Cuz if you like it then you shoulda put a frame on it/…/You'll be glad once you've coopted it/If you liked it then you shoulda put a frame on it," and defiant about needing "no permission," her lyrics continue to call on the listener to "mix memes":

> The subvert is what I prefer,
> To swerve, to make these words
> Here's a way that makes memes by making memes
> And delivers me to a festering indemnity for aeons
> Pull memes into your swarming arms
> Say I am what you own
> Let it be intertextually grown, iterably sown.

Ultimately, Karasick's piece is almost a total subversion of Beyoncé's dance tune. Radical language, countercultural disturbances, and deconstructive acts become the weapons of choice in the cultural war. And Karasick's lyrics are not the only way she applies her theory. The video itself is transformed through remixing pre-existing materials and through a patch: Karasick's face replaces Beyoncé's. Images of a nuclear bomb, veiled women, rabbis, orthodox Jews, queer icons (Gertrude Stein and Alice B. Toklas, for two), and many philosophers (Hannah Arendt, Ludwig Wittegenstein, etc.) are inserted around, behind, and over the original video graphics of three Black, beautifully choreographed, scantily leotarded singers and dancers.

Karasick does not ignore color politics as she transforms and deviates from the original text either. Beyoncé is African-American. Karasick is of Jewish descent. It would be a simple act of negation to merely erase Beyoncé, and Karasick is far too mindful for that. Instead, along with her own face, she inserts Jewish iconic imagery, significant works from art history, Hebrew text, symbols from the Kabbalah, images of coopted consumers, and sexy women in nontraditional roles – from bodybuilders to pole dancers to fire breathers – interacting with the conceptual framework of language. Even Gertrude and Alice are holding a sign calling upon viewers to "Adopt a syntagm program." Syntagmatic analysis is derived from the science of semiotics. In the realm of syntagmatic relations, the signifiers are material objects – sounds, words, images – that occupy the planes of time (sequence) and space. The signified is the concept – the frame that Karasick refers to – of conceptual ideas or the perceptual leap in the act of glossing an "ellipse, full of textual slips." Thus Karasick is urging us to bring conceptual analysis to bear on the media including language that surround us in our world. Where Beyoncé's 'acting up' was about putting on a new pair of designer jeans and sporting a new "man that makes me, then takes me/And delivers me to a destiny," Karasick's subversion of the original is a splintering of perspectives to subvert issues related to language, gender, race, and religion, and to *détourne* a new space for political critique and conversation. Her video is successful because what she advocates is a plan of action, that is to say it is a conceptual *process*, not a product.

Memes can work in the reverse direction too, with digital memes materializing as contagious references in the real world too. A significant example of this is *Catt,* a fake sculpture drawn from a digital meme and exhibited in a real gallery (2010). Pioneering New York-based net.artists Eva and Franco Mattes (a.k.a. 0100101110101101.org) have a history of performative pranks and the artistic re-creation of events online (see the discussion of *Thirteen Most Beautiful Avatars* and "7,000 Oaks" later in this chapter) and off. They are known for their hacks of feature films, advertising, online technologies, and video games. Disturbing game hacks include *Freedom* (2010), in which the Mattes try to engage MMORPG players in conversations about what they call another "alternate reality," contemporary art. All negotiations were in vain and the 'artist' in the game was gunned down after only a few words (http://www.aksioma.org/systsem/); and *My Generation* (2010), which culls found Web footage of game players who lose control after game losses. The Italian group's most infamous prank was when Franco Mattes staged an Internet 'suicide' by hanging and then recorded the unscripted responses of visitors to the live webcam footage. Many were horrified, but – looking through the numbing veil of spectacular culture – just as many visitors were indifferent. The Mattes' work *Catt (Fake Cattelan Sculpture)* takes memetics to a new level. The sculpture consists of a taxidermed cat and bird. The grey tabby is inside a bird cage, looking very disgruntled, and a budgerigar is perched on top of the cage looking down at the cat. The concept itself is a re-creation of a memetic, digital image found on the Web. The original was a LOLcats meme posted at the Icanhascheezburger.com site under the caption "epic fail" (http://icanhascheezburger.files.wordpress.com/2007/11/funny-pictures-bird-cat-cage.jpg). As an art prank, the Mattes created, framed, and presented this sculpture in the style and manner of a prominent Italian artist celebrated in gallery circles, Maurizio Cattelan. The imitated artist is known for his remakes of other artists' work, a self-ironizing comic flare and satirical works; the Mattes therefore submitted this work, *Catt* in-the-style-of-Cattelan, to a group show called *Weasel* at the Inman Gallery Annex in Houston, Texas (2011). After many rave reviews for Cattelan's mastery, the Mattes made their intervention known. Upon learning of the hoax, the co-curator Kurt Mueller explains, "I would expect an exhibition titled *Weasel* would make viewers suspicious – and some were – but even the sharpest eye is influenced by the way something is presented. We're all prey to context." His curatorial partner Chelsea Beck added, "Sometimes 'the label' is as important as the content" (Inman Gallery, Weasel Press Release). The apologia of a press release quotes the Mattes' explanation about their act of fraudulent 'authorship' and re/possession of a stolen digital meme. "I was never very good at making art. But, modestly speaking, I'm pretty good at copying," declared Franco Mattes (with tongue securely planted in cheek), "so I thought, let's wear the mask of a famous artist and see what happens: Will people realize it immediately? Will they say 'this is his worst work ever?' Or will they love it?" These artists have a long history of impersonation and 'faux' creations, fraudulent names, and fictional personae. For instance, the duo invented an artist for the Venice Biennale one year and have "remixed works of other Internet artists and stole[n] parts of many dozens of different art masterpieces, from a Kandinsky to a Warhol, in a performance that lasted two years." Eva Mattes situates these *détournments* of originality in the realm of appropriation: "'Appropriation of visual culture…has always been part of human history, of the way humans progress and learn, and art is no exception. It would be more useful to look for references in a work of art instead of originality.'" The Situationists would have approved of these methods. According to art historian Inke Arnes, "the (mis)appropriation and repurposing of conventions produce shifts in social consciousness" that are at the heart of *détournement* (quoted in Shanken, 2009, 30).

The second strategy the Situationists advocate is the *dérive,* an experimental mode of behaviour "linked to the conditions of urban society": "a technique in real time of rapid passage through varied ambiances" ("Theory of the *Dérive*"). Drifting without purpose allows an awareness of the environment. It is a tactic for remaining or reclaiming embodiment outside of the media spectacle. This is especially important as specular power becomes wholly reintegrated (as Debord said it had in 1988; Harold, 2007, 14), and political resistance is branded and sold back to consumers. Classic *dérive* uses a person's psychological awareness of a neighborhood to navigate it in novel ways – just like practitioners of Parkour practice illicit explorations of urban space. Fashioned as a replacement for work and entertainment, Harold says that drifting was both a material and rhetorical

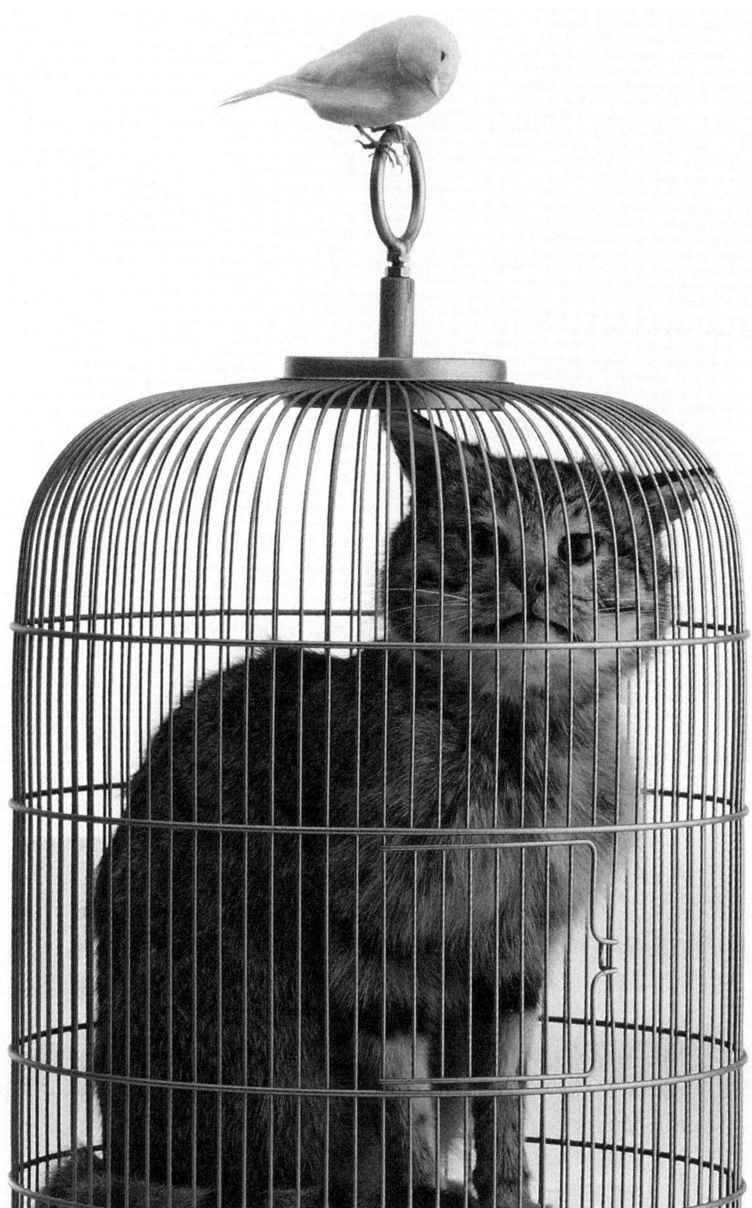

Figure 1 Eva and Franco Mattes, *Catt*

practice that sought to redefine the self as an agent through imaginative uses of space (Harold, 2007, 14–15). At its best, it was a game of chance; it fostered a new kind of ambient embodied spectatorship outside of the sphere of commercial exchange. Rooted in the psychogeographic use of space, the body became a rhetorical agent of change through creating a procedural aesthetic writing with the body in space. (Harold, 2007, 16). The *dérive* is an exploration of information space, a mapping of currents in cities, of dataflows, fixed points (nodes), and vortexes in urban space. The purpose of the *dérive* is to discover "blockages, paths and possible fissures" (Harold, 2007, 16). "[R]eification is never complete," Debord says (quoted in Harold, 2007, 16); the abstraction of Situationist resistance must always remain impenetrable to language and to logic. This interrupting mechanism is always already accompanied by repetition and stoppage. DIY culture as interruption is a way of interfering with the spectacle to make it visible, to make rhetorical choices that are situated. *Dérive* makes use of the tools at hand, says Harold (2007, 17), and through the conjoining of parody and deadly serious political content the contradictory nature of our times is reflected back (Situationist International).

Along these same lines, Michel de Certeau advocated the practices of everyday life (in his book of the same name) as a mode of resistance to political domination. In the everyday he saw two kinds of practices: strategies and tactics. Strategies, which are the purview of those in official situated positions within the political infrastructure, allow for the creation of social relations and the engineering of social realities. In short, strategies are institutional forms of control. Tactics are the methods of the masses. Contrary to expectations though, de Certeau does not construe the masses as powerless – quite the opposite. As active agents, the individual's practices of consumption are a continuous disruption in the reality projected by official sources. Tactics are detours or short cuts that we use to navigate around official power nets. The very possibility and hope of disrupting the flows of official power are implicit in the tactics of everyday practices. Subversive tactics are, by definition, fluid and unmappable. De Certeau's network is designed for hackability, remixability, and mashability, like Karasick's detour through Beyoncé's pop culture framework, or like Web 2.0. Social media seem to offer tactics for resistance, but this is undercut by the capitalist drive behind the forms. If users are willing to relinquish control over their creations by uploading them to social media sites in exchange for large amounts of storage and the ability to customize the interface, then the user becomes part of a community of like-minded souls. If you upload your material to YouTube, for instance, Google owns the rights in exchange for giving you access to its potential audience. This sum is greater than the parts, though, since the conversations that develop and the networks that are forged often cross different social media platforms and modes of discourse. For example, transmedia storytelling and fervent fan cultures resulted in the resurrection due to popular demand of two different television franchises: *Star Trek* and *Battlestar Galactica*. In both cases, the new versions far surpassed the old in terms of quality and staying power. In the case of the former, some diehard fan fiction writers actually became the next generation of writers for *The Next Generation*. Marketing campaigns can also go viral, as in the case of the British company Dubit Insider, and can coopt social media users back into consumerist activities. It recruits 13- to 24-year-olds as "brand ambassadors" to do its marketing for it. In exchange for company incentives, these "peer leaders" use so-called 'word-of-mouse' technology to post and forward messages promoting their products via email, IMs, on social networking sites, at events, and through home parties. The capitalist impulse thereby translates teenagers into spambots working on the company's behalf. In all of these cases, the content creator becomes grist in the capitalist mill. These are the antithesis of de Certeau's network of hacker activists.

Consider Katherine Sweetman and The Infinity Lab's hack of the Los Angeles-based Getty Museum's "Video Revolutionaries" site in 2008. In an apparent attempt to be *au courant*, the Getty created videorevolutionaries.com to encourage "emerging artists 'to be a part of the video revolution'" at the California Video exhibit (theinfinitylab.com). Participants were promised a public screening of the top-ranked entries. Ostensibly an open contest – designed to be "Part YouTube, part 'American Idol'" (theinfinitylab.com) – entrants actually had to submit to a rigorous screening process to be pre-approved by the Getty's judges before they could submit work. In protest at the closed nature of the contest, the Infinity Lab launched what they called "a digital hijack" which resulted in their videos being ranked the most viewed and highest rated in the competition. The 'revolutionary acts' carried out in Sweetman and the Lab's videos are banal activities by tinfoil-masked perpetrators next to a child's backyard swimming pool with a toy skeleton. Highbrow aesthetics and lowbrow culture do not mix very well when old rules are applied to new practices and schools of thought. In the end, the Getty did show the Infinity Lab's videos, but in a subterranean room spatially removed from the rest of the screening. At the Infinity Lab's website, they claim: "If a revolution is a modification of an existing methodology, constitution, or structure, then The Infinity Lab has succeeded in a tiny revolution that is not 'homogenized' by The Getty or the videorevolutionaries.com website" (theinfinitylab.com).

The person in de Certeau's network was neither a voyeur, a spectator, nor an author. She too was a hacker or hijacker who used fragments, trajectories, and alterations of space, including material outside of language and representation, to try to find an outside to the system of representation, as in Electronic Disturbance Theatre's *FloodNet*. The drifter's path is directionless and unpredictable, and therefore unquantifiable within the Spectacular net, outside of surveillance. Disorientation is necessary to reorientation (Harold, 2007, 17).

Psychogeography is also a hack. Christine Harold argues that without a predefined goal, it is a purely oppositional strategy. The importance of its lack of goals means that it cannot be diverted because it exists only in moment-by-moment movements like flash mobs. It is the act of mapping the city, and its information spaces. It allows for consumption of the city without being consumer/subsumed/tainted by it (Harold). It was a tactile exploration of the city along spatial rather than temporal coordinates. It is not an end to history, as Harold claims (2007, 19), but a celebration of immersion in the timeless space of the present moment. It is immersion in the non linear information space of the map. Unofficial tactics may be inefficient but personal, aesthetically pleasing routes and detours, aimless, meandering, refusing the economy of exchange.

Back in 1944, Theodor Adorno and Max Horkheimer in "The Culture Industry: Enlightenment as Mass Deception" had warned that mass production would inevitably lead to the coopting and standardization of consumers themselves. Where traditional Marxism had warned that capitalism would impoverish the masses, Adorno and the Frankfurt School saw the ever-increasing abundance of goods fostering an impoverishment of a cultural nature instead. After Adorno and Horkheimer fled the Nazi regime, they landed in California, which for highbrow critics must have been about as close to hell as they thought they could get. In motion pictures, they saw nothing but the equivalent of capitalist propaganda as commodified emotion. Consumerism is designed to satisfy needs – even if they are needs that the person never knew they had – and so there is no outside to the system. The more goods a person has, the more they need. The mechanization of culture is designed to meet a big audience's needs, rendering shows like *American Idol* or *Britain's Got Talent* as addictive merchandise to be consumed. This was what Walter Benjamin (also a member of the Frankfurt School) had warned about in the loss of the aura of the original work of art; under capitalism it is commodified, boxed, advertised, and sold. It was in the collision of highbrow and lowbrow forms that the elitist Adorno and Horkheimer saw the greatest danger, for, they (quite rightly) feared the downfall of the pinnacle of civilization in European classical music being degraded by hobnobbing with popular songs. Imagine how surprised they would have been when social media in turn became a device for recruiting pop culture-hungry under-50-year-olds back into the fold of classical music, as the Pacific and Los Angeles symphonies are doing in 2011 (Berger, 2011). Jean Baudrillard also foresaw the seeds planted for the disappearance of art in the rise of the production of culture. This is what he called the "Xerox degree" of culture, where mass repetition drowns out art:

"one is the other's vanishing point, and absolute simulation" (1988, 173). Conceptualism has not changed very much in that regard as the new digital conceptualists too try to foster an "outside" to Spectacular culture. Hackers have not appropriated de Certeau's methods as such, but have adapted and re-created them in a new DIY media. Popular culture now uses these hackers as celebrated countercultural anti-heroes, as in *The Matrix* (Wachowski, 1999),[9] while artists like the Mattes, Scott Kildall, and Cao Fei use Second Life and other venues to interrogate the nature of virtual identity outside of specular culture. In our mass-produced age, we all build selves, identities out of customizable and remixable prefabricated elements. The Mattes, Kildall, and Cao take the self to new levels of artistry.

As a filmmaker, Debord sought a process of making images and fostering revolutionary fervor outside of representation. Agamben identifies the revolutionary element in Debord's filmmaking strategies as being the act of breaking the frame, which interrupts the flow of the medium. Philosopher Gilles Deleuze also theorized that, in cinema, every act of creation is also an act of resistance. Debord was not interested in resistance (which he feared was merely an act of negation), but in the disruption of the spectacle to make the medium visible. Debord strives to make the image disappear in the films he creates through the use of both all-black and all-white screens. These are a kind of nonrepresentational image of, respectively, an image of all images and the image of none. All and nothing. Agamben argues that Debord's film *Society of the Spectacle* also incorporates a loop so that the repetition is complete, enfolding the stoppage as a process *ad infinitum* within itself. In our digital time, Rita Raley defines the form of art-activism known as Tactical Media in her book of the same name as signifying "the intervention and description of a dominant semiotic regime, the temporary creation of a situation in which signs, messages, and narratives are set into play and critical thinking becomes possible" (Raley, 2009, 6). This is the interruption, the eruption, and the breaking of the frame that we see in the best remixed digital media.

The value of remixed media in the social network is in participation. The ultimate mark of success is when a video goes viral, performing its message through its audiences as it circulates, accruing versions, adaptations, and parodies. Critic Jean Burgess says,

> [I]t is necessary to see videos as carriers for ideas that are taken up in practice within social networks, not as discrete 'texts' that are 'consumed' by isolated individuals or unwitting masses – a 'copy the instructions', rather than 'copy the product' model of replication and variation. These ideas are propagated by being taken up and used in new works, in new

ways, and therefore are transformed on each iteration; and this process takes place within and with reference to particular social networks or subcultures (Burgess, 2008, 108).

Semiotics and structuralism are both rule-bound as well and set out to reveal new systems that are governed by laws and codes. They therefore adapt themselves very well to algorithms in the digital age. This new system functions like a language (Aaron, 2007, 6). Philosopher Louis Althusser starts from the position that cinema is ideological, and explores how, working ideologically, film comes to construct networked subjects – that is, subjects that exist within a network of 'imagined relations'. By blending psychology with the effects of material and social conditions of the medium, Althusser's thinking is revelatory. Apparatus theory in turn sprang from his thinking and the camera/projector binary began to incorporate the spectator as a part of a new tripart understanding of how film works. To the spectator, film endows the spectator with the illusion of authorship, according to Michelle Aaron: "the film apparatus produces the subject and positions the spectator as a false author of the image" (Aaron, 2007, 14). This is a virtual space that the subject/spectator occupies. Roland Barthes further enriches this formula by exploring the readerly/writerly relationship as what he calls an "'infinite intercourse of codes'" which "renders the reader *a site* through which the controlling systems of conventional thought shape the meaning 'he' generates" (Aaron, 2007, 17–18; emphasis in original). The spectator is a voyeur to be sure, but shows little evidence of authorship in this configuration. Protocol, both vertical and horizontal, is what really governs.

Take, for example, Sina Weibo. Internet censorship is a fact of life in China, earning it the nickname of the Great Firewall of China.[10] Facebook is naturally blocked as well,[11] but Sina Weibo is a very similar popular Chinese social networking site that is overseen and closely regulated by the government. It was launched in 2009. Weibo is the name for microblogging in general, so while Sina Weibo means Chinese microblogging, Facebook is so high profile that it is thought of as the Chinese Facebook. Other microblogging sites exist as well, but Sina Weibo is the most popular, drawing approximately 140 million of China's 195 million social networkers. An ad that Weibo ran to promote its own service is a satirical remix and mashup of a number of Western sources of material. In fact, the ad demonstrates an intimate and current knowledge with what is going on in the world, especially in popular culture. Called "500 Million Facebook Friends to Sina Weibo," the ad starts with a narrative explication of the history of Facebook. It has been said, it tells us, that 500 million men and women who are followers of Facebook are chatting worldwide. The initiator of this communication revolution is

Mark Zuckerberg. The ad then switches to Zuckerberg's own Facebook page where he posts an alert that all of his 500 million friends are gone. The voiceover is in Mandarin (the typed chat is in English, with terrible spelling and worse syntax), but the meaning is clear. He is particularly concerned with his hot girlfriends and their absence as he chats with his few remaining friends (including Tyler Winklehoss who tells him that he has a PR problem and people just don't like him – clearly a reference to the 2010 film, *The Social Network*). He discovers that all of his friends have "flowed" to China, switching to Sina Weibo. The ad then jump cuts to a scene from Christopher Nolan's 2010 film *Inception.* Ariadne, the character of the architect of the virtual world who is played by Ellen Page, says, "We are going to create a new social media space. We have to create a new world. From now on, we don't need to jump over the wall because we are going to have Sina Weibo." The wall, of course, is a reference to the firewall. She then speaks to Cobb (played by Leonardo Dicaprio) and tells him in Chinese, "There is no such thing as a revolution without a sacrifice or a death. In a real revolution, you must pay with lives. If we could have gone on Weibo, it would have been better." The picture breaks up into jigsaw puzzle pieces revealing a scarf marked with Sino Weibo draped around, in succession, the Eiffel Tower, the Statue of Liberty, the Fernsehterm Berlin communications tower in Alexanderplatz, and L'Arc de Triomphe – all powerful symbols of freedom in the West. Next, the ad jump cuts to American President Barack Obama saying, "Yes, Weibo can." Muammar Ghaddafi appears and says to the press, "Yes, it would have been great if we had been on Weibo earlier." Osama Bin Laden sends a pirate broadcast where he explains that due to the poor datastream qualities this is his first appearance in three years, and his first time on Weibo. Bill Gates says that Microsoft is no longer soft (i.e., weak or not strong enough) anymore because the company is going to be on Weibo. Cobb/Dicaprio from *Inception* says to Ariadne/Page that he saw her on Weibo, and asks her to add him as a friend so that they can talk to each other. Next a clip of *Love and Other Drugs* (Zwick, 2010) plays. A naked Anne Hathaway (playing character Maggie Murdock) sits on the edge of a bed talking to an equally naked Jamie Randall (Jake Gyllenhaal). (Nudity is rare in films shown in China and often subject to censorship.) She says to him, "We have 2,000 more friends now." "Wow," he says. "So fast." The next clip, from a period Asian film (which I have not been able to identify, but which seems to depict an event from the Chinese Cultural Revolution), shows an overturned bus in a river and a man standing on top of the bus. He yells at the crowd, "All the people are within my grasp now. Follow me!" The byline tells viewers that foreigners from outside China can now use Weibo.com's services (English language – which launched in June 2011). The final scene reverts to Mark Zuckerberg's own status page where he asks, "Is there a sister (i.e., a young, single woman) who can teach me Mandarin?" What is remarkable about this ad, beyond the skillful remixing and mashup skills and its astute awareness of up-to-the-minute Western popular culture, is the way in which it translates the language of the Chinese Culture Revolution into a call to arms to join a revolutionary movement in social networking. It is rare for Chinese media to wield irony so effectively and rarer still for such subtleties in politics, past and present, to be addressed so openly. More interesting still is the fact that only 20 films from the West are given permission to be screened in China each year,[12] but the reality is that, banned or not, virtually everything is available. This currency with the West is seen in this ad, which displays a deep fluency with its media, social or otherwise. It is shocking to see revolutionary language intertwined with notions of a virtual revolution in a distinctly Chinese context. This underlines very clearly how irrelevant revolution has become in the age of the spectacle.

The Situationist International's revolution in May 1968 proved most emphatically with the silent reinstatement of the French government that real revolution was no longer possible then either. The arrival of the information society on the rising tide of multinational capital was foreseen by curator and dealer Seth Siegelaub as the fulcrumb "between the new economics of aesthetic value and the politicized cultural critique that erupted in the late 1960s" (Alberro, 2003, 3). There is a paradox, of course, involved in the exhibition of conceptual art, since "the egalitarian pursuit of publicness and the emancipation from traditational forms of artistic value" were at odds with the space of the gallery as a vehicle for "advertising and display" (Alberro, 2003, 5).

As the most revolutionary art form of the Twentieth century, conceptual art ironically achieved its power through a move away from the visual and by embracing dematerialization. Marshall McLuhan had similarly seen advertising as the art form of the Twentieth century. Like advertising, Alberro states, conceptual arts' use-value is located in its notoriety, its cultural caché, its advertising (Alberro, 2003, 131). Without an aesthetic frame to attach its reputation to, the work erases the distinction between high and low culture, and shifts the focus instead on to the work's *documentation* (Alberro, 2003, 131). Where pop art merely repeated cultural icons, conceptual art programmed them within the world to break down all useful distinctions (Alberro, 2003, 131). It was in the gap between the absence of representation and the withering of the art object that conceptual art also made the political structure of the gallery visible in the late 1960s.

One of Siegelaub's celebrated shows was *The Xerox Book* in 1968. Perhaps inspired by McLuhan and Fiore's 1967 celebration of xerography's leveling of the playing field by giving everyone access to the tools of production, the show demonstrated how irrelevant copyright had become. It showcased the ability of the technology to steal from other authors.[13] Siegelaub explored authorship as a practice open to everyone foregrounded in the medium of communication as a part of the public domain, and as a frame ripe for artistic exploration. More remarkably, anyone who bought the book owned the work, as ownership was conveyed to the public through exchange (Alberro, 2006, 147). Alberro deems the show spatial – running for 175 pages – rather than temporal – running for three weeks (Alberro, 2006). Each artist in the show submitted 25 pages of what we can now read as art algorithms or algorithmic code.

> the mechanical 'impersonal nature' of xerography depersonalized the production process, negating the skilled hand of the artist in a way that once again resonated with the visionary writings of McLuhan and Fiore, who observed that "as new technologies come into play, people are less and less convinced of the importance of self-expression." But the feature of the photocopy medium that most interested Siegelaub was its negation of the aesthetic component (Alberro, 2003, 135).

The extremely low quality of the reproduction is noise that further eliminates the visual (although cost ultimately forced Siegelaub to print the *Xerox Book* by conventional printing means rather than photocopying it). This becomes meta-art, self-reflexive art, art concerned with its own technologies of production – similar to much digital art that will come after. Like code, language becomes a means of real-time information transmission, as color had been before it (Alberro, 2003, 140). Sol Lewitt comes closest to this with his pages that give command code-type instructions for procedural engagement (Alberro, 2003, 142). Lewitt's piece was participatory, and designed to be executed by users. Joseph Korsuth's contribution was perhaps the most remarkable. It was pure documentation, recording all the constituent parts of the catalog:

> Kosuth's project matter-of-factly itemized the constituent elements employed in the production of "The Xerox Book": "Title of the project," "photograph of the Xerox machine used," "Xerox machine's specifications," "photograph of collation machine used," "collation machine specifications," and so. On. The literality of the work echoed LeWitt's dictum that art should not instruct the viewer but should self-reflexively present information (Alberro, 2003, 147–8).

The implications for documentation becoming or replacing the work of art were raised for the first time in the context of this show. Art was billed as one form of documentation. More and more the conceptualists moved into rule-based art, which we can easily be seen as a precursor to computer algorithms and protocols. As some examples from the 1960s reveal, On Kawara began on January 4, 1966 to paint every day's date on a monochrome canvas (in one of eight sizes). Other rules guided his art too, such as it being sealed in a box with a newspaper clipping from that day or destroyed if not finished by midnight. Jasper Johns used grids with alphabets and Warhol made multiple copies of copies and series of copies. Ed Ruscha photographed "Every Building on Sunset Strip" at high noon from the same distance away (Cohen, 2005). Under digital rule-bound art, art has become much more about visualization and about data. Swiss artist Ursus Wehrli, for instance, works on pieces like *Pattern Making*, that foreground visual order and statistical recombinations. I will talk more about the importance of protocol later in this chapter. The Australian hacker-artist-linguist Mez is queen of documentation because, in essence, her work aspires to pure documentation. I will discuss her netwurks in Part Three. Documentation as art reaches its logical conclusion – or absurd limit if you prefer – with Miami artist Manuel Palou's "'*5 Million Dollars 1 Terabyte*': Pirated Data Displayed As Art" (2011).

In an age of digital media, Debord's *strategies* have been replaced by socially mediated aesthetic events or hacks – what Raley calls *disturbances*.[14] Tactical media are the 21st century's answer to activism after the Twentieth century demonstrated so clearly that protest and social change merely replace one kind of power dynamic with another. Tactical media are a temporary glitch, a noise, or interference (Raley, 2009, 28). They function as "a tool for creating temporary consensus zones based on unexpected alliances" (Raley, 2009, 6).[15] In the absence of the possibility of revolution or rebellion, tactical media stand out most remarkably for their impermanence. They are not essentialist or oppositional, but instead are drifts: performative, improvisational, collective, random, speculative, temporal, locative, ephemeral, unexpected, open-ended, often speedy and generative. They are tools of protest and assembly that use other digital tools to disrupt media, lines of power, and multinational capital. We could read them as systems art that seeks to investigate the process of an art form. They may be our last great hope for resistance, creative critical engagement, and protest in a Spectacular age where all our media are integrated into our lives in ways that would have been unimaginable to Debord.

Disturbance (action + event)

Art is not a notion, but a motion. It is not important what art is, but what it does (Gilles Deleuze).

"Over and over," McLuhan explained in a letter, "I've talked to groups and individuals about new technology as new environment. Content of new environment is old environment. The new environment is always invisible. Only the content *shows*, and yet only the environment is really *active* as the shaping force" (McLuhan, 1987, 311). The endless repetition of familiar content in a digital age is what makes the environment visible and fosters social networks. Where lawmakers would call this process of reusing familiar images piracy, we can see this as a social critique, using the vernacular of our day. It is the very fact of repetition that fosters a conversation, and generally one in the same medium as the original. Performance art, which is simultaneously a new environment and a technology, has this active nature. As a break boundary, the systems art of McLuhan's performance art or Debord's urban remixes and wanderings contain within them several older media, including the body and performance. In television and film environments, the body and performance are mediated through the introduction of a host of technologies to a point where, according to Philip Auslander, they can no longer be live because even live performance is now a mediated event. In fact, there may be no outside to our mediated culture at all any more. Informed and mediated by broadcast technologies, so-called live TV, TiVo, and other providers allow broadcast to be paused or edited in real time. We can see the World Trade Center Towers collapse again and again, for instance. But while these events unfurl before the lens, the spectacle of terrorism becomes the terrorism of the spectacle (Baudrillard, 2002, 30), and this brings into sharp relief the fact that the mass media have changed forever the notion of live performance. Authenticity in the heyday of the written word was a shared experience. In the Twenty-first century, authenticity is only ever possible when embodied and it is something that we must experience live through our own technology, the medium of our body.

Broadcast *mass media,* which Debord saw as so dangerous, was about controlling a mass audience, the mediated reproduction of an event, and repeatability. Instead he invested his energies in aesthetics, which were focused on disturbance, action, and event. As conceptual artists, McLuhan and Debord both contributed to the philosophy around the dematerialization of the *objet d'art* and to a new emphasis on an "art and aesthetics in which ideas and discourse – not the formal conventions of the medium – constitute the principal elements" (Lillemose, 2006, 118). The third most important thinker on art, technology, and conceptual art in the 1960s was critic Jack Burnham. Burnham saw the rise of conceptualism and the accompanying dematerialization of the work of art as fallout from the rising tide of ideas by media theorists like McLuhan, and "as an indicator of a 'transition from an object-oriented to a systems-oriented culture' where 'change emanates, not from *things,* but from *the way things are done'*" (quoted in Lillemose, 2006, 124). This was at the same time that McLuhan and Debord foregrounded live experience as the most effective system to express resistance and get things done.

Tactical media: public disturbance after the decline and fall of activism

Activism has gone the way of creativity. Where the material nature of art was predicated on socio-economic and social forces as a contextualized form, dematerialized conceptualism was political not on account of its form, but on account of how and where it was situated. Conceptual art was soft or fluid, and, according to Jack Burnham, software was a metaphor for art (Shanken, 1998). Burnham foresaw that the future belonged to models of social interaction (Lillemose, 2006, 127), and social media fit that bill perfectly.

Web 2.0 culture has wrought a revolution in communication, making it possible to broadcast ideas to groups very easily. Social networking including blogs, Google+, and Twitter makes your content searchable. An extension of the open source, open software, Creative Commons, and open content movements, they foster peer-2-peer or site-2-site conversations through foregrounding social networking; Facebook, for instance, operates on the basis of 'friends.' Social media provide unique and stable URLs, or what are considered permanent links (for the next five years at least), and are syndicated, giving users the opportunity to subscribe. We have seen Web 2.0 culture and how blogs, for instance, in particular have transformed journalism, publishing, and news dissemination as major daily after major daily have teetered and fallen from their pedestals. The participatory content of Web 2.0 materials is frequently dismissed by their critics as so much drivel from so-called cut, copy, and paste artists, sneered at as "opinion indicators" and "pyjama journalists" (Lovink, 2006). Despite many claims of their revolutionary nature, blogs do not shatter the wall of the Spectacle. What they do do is create a gap, a social space, a fissure for the reinsertion of DIY networked media. Web 2.0 sites like YouTube, Facebook. blogs are profound tools of not political but *personal transformation*, just like many other writing technologies:

> "by keeping a daily record of their rites of passage, bloggers often give a shape and meaning to the stages and cycles of their lives that would otherwise be missed in the helter-skelter of modern existence." Foucault scholars would say something similar, namely that blogs are "technologies of the self" (Lovink, 2006).

Like the typewriter though, which McLuhan called a tool for transcribing speech, they operate from within a political system. While they might foster personal transformation, they will never forge political change on their own. Instead they can only mark and stand as a measure of political or structural shifts within the status quo. They give the illusion of control while simultaneously being more and more tightly boxed behind corporate walls. While Web 2.0 technologies are tools, they are in their essence a way of starting and maintaining a conversation first and foremost with our self and then, later, with others. Their value lies not in their content as such, but in what Henry Jenkins calls their "spreadability" (2009). Participation is what creates stickiness for users (Jenkins). DIY culture in particular is an attitude more than a technology (Lovink, 2006, xii). The purpose of the blog is the creation of a social network that crystallizes around a particular content. Blogging is more akin to an oral form than it is to print, and like oral forms, finds a footing in repetition. Geert Lovink observes that:

> "The whole blogosphere is an example of how transcending the top-down hierarchical models of old-media technology with new-media technology

releases diversity and new voices and creations." Against this commonly held view that diversity is a good thing, we can hold the loss that comes with the disappearance of familiarity and common references. Blogging alone . . . is a social reality which cannot easily be dismissed. Most blogging is what Bernard Siegert calls "ghost communication" (Lovink, 2006).

Like Debord's ideal plagiarism, many 2.0 sites are primary spaces for the sharing of information among like-minded users. YouTube is an exemplary model of social media that fosters a conversation and community formation through an exchange of remixed and mashup videos. "Successful 'viral' videos have textual 'hooks' or key signifiers, which cannot be identified in advance (even, or especially by their authors) but only after the fact, when they have become prominent via being selected a number of times for repetition" (Burgess, 2008, 105). They are often silly little repetitions of a formula like the viral series of videos "Where the Hell is Matt?" It was one of the earliest Internet memes to go viral and the one considered the most successful viral video to date. The original 2004 video[1] shows Matt Harding, a former computer game designer, dancing a goofy dance in stunning and/or iconic locations all over the globe. At his website, Matt tells us that he created the first video himself, but his second two trips around the world in 2006 and 2007 were financed by Stride Gum. The 2007 trip became a social media phenomenon – with more than 38 million viewers to date – as people could follow Matt via his website and know in advance where he was going to be in order to join in. With the addition of this participatory component, the videos start to become more interesting as the exuberance and *joie de vivre* multiplied. The *New York Times* called it "an almost perfect piece of Internet art" (McGrath, 2008). *TIME Magazine* named it the best viral video of the year (2008). NASA selected it as "Astronomical Photo of the Day" – even though, as Matt points out, it is neither a photo nor celestial. But the story does not end there, since, with the arrival of all of the accolades, Matt finally divulges the truth: the video is a hoax. Nothing is real. He had never traveled outside of the U.S.A. in his life. All the people who joined him were actually anamatronic puppets. All of the images of foreign places were photoshopped around footage of him in front of a green screen. He is not a programmer, but an actor hired by a viral marketing agency in New York called Buzz!Brain. The productions were not a homemade DIY effort, but professionally produced and extremely costly. Spin-off commercials have since been financed by Visa and the World Cup. There is a book and you can buy the T-shirts for the 2011 tour. It was the ultimate fake or fraud, a mere script, a plagiarism of someone else's life that perhaps somewhere deep down we would like to live. It is perhaps one symptom of the death of authenticity, the devolution and devaluation of authorship, and the rise of 'personal' narratives (even if a corporation writes them for us). It is also emblematic of the stickiness of a particular kind of Web aesthetic that eschews a complex narrative with characters and a plot, and favors instead the interruptions of many little repetitive clips, the performance of something simple and catchy, and capture as a double-pronged mode of discourse and contagion. As a Western corporate product, however, what "Where the Hell is Matt?" lacks is something that is the reason for creating in tactical media: a social conscience and a political or hacktivist agenda. In reference to literature, Jacques Dérrida refers to the *démultiplication* of postmodernism: the proliferation of voices, channels, and modes in our contemporary time. The best new media art captures this dialectic or chorus within it and speaks in multiple tongues and modalities as a means of performing its subversive content.

In Part Two of this work, I will discuss the queering of mainstream media as one form of subversive dialog to mass culture and a way of opening a space for different kinds of creative practices and different kinds of narratives. These subversive ghosts haunt the content of the form. Presences that have stepped in to make content king on the Web, but also to remind us in a space, that is generally deemed to have no history of its own, of the already forgotten forebearers of revolutionary goals, especially of feminist media activists who have used DIY culture for years to organize. Back in the 1980s they set up bulletin boards and online communities or practiced media strategies and tactics. These include groups like VNS Matrix, a collective out of Australian who were among the first to coin the term *cyberfeminism* and The Old Boys' Network in Germany, a cyberfeminist group who strive to open windows of access for women in high-tech careers.

A brief history of feminist activism in electronic spaces will help to illuminate how *détournement* and *dérive* can be used for political ends. Before there was a so-called Twitter revolution in Iran in 2009, there was Chernobyl in the Ukraine in 1986 and Z-Netz, which were German-networked BBSs. Before the World Wide Web, there was something called the Internet that was accessible by dial-up and hosted communities most often on bulletin board systems. One of the first and the most famous of these was the WELL. The WELL was so celebrated in part because it was one of the few spaces where everyone could have a voice. That was very different from groups like CompuServe and AOL. Women did not attend the WELL in equal numbers as men, but, unlike elsewhere, sexual harassment was not tolerated. German computer artist and activist Rena Tangens was a pioneer in online media in the 1980s with her //BIONIC BBS. Tired of being interviewed by cybersexed, cybersleazy, cyber-harassers who would print serious articles with cutsey female models (Waltz, 2005, 93) what Tangens really wanted was to put women behind the wheel of their own destiny. The problem was not getting women to be present; it was getting them to speak up. She argues that the first question hactivists must ask is "whose biology and psychology is being programmed into their machines? ... Is there a gender bias to the choices made?" (Waltz, 2005, 93). She set out to design systems that fostered collaborative work rather than "replicating existing power relations" (2005, 93). Many systems at the time assumed that you were male unless you identified yourself as something different. The disaster at Chernobyl became a different catalyst that encouraged male hackers to put aside other methods to work with Tangens' system. The nuclear accident in the then Soviet Union was covered up and no information was forthcoming from official channels. There were rumors of a terrible disaster, but nothing was substantiated. Tangens' Atari Computer, which she used to manage her art gallery mailing list, allowed her to hack official barriers. At first she tried to use FidoNet which "allowed the whole world of Fido-using computers to exchange files and messages, bouncing them over modems ... until all reached their eventual destinations" (Waltz, 2005, 95). Bizarrely, FidoNet's owner, hacker Tom Jennings, saw what Tangens was trying to do as terrorism and barred her from the system. So, instead, she worked with other hackers to create the Zerberus mailbox, a feminist BBS, and Z-Netz, its network. "The result was the rapid and wide dissemination of the real Chernobyl news across Europe and then the world" (Waltz, 2005, 96). Once the crisis was over, they did not disband, but instead became a permanent part of the European high-tech scene, fighting the war in the Balkans and political censorship in Germany, for example. Z-Netz's privacy function is built in and it has anti-harassment features, which automate your ability to reply to offensive idiots with standard form responses like "'Your message with subject <blah-blah> has been ignored by <your name>" (Waltz, 2005, 96). Tangens' art installations similarly require full participation and build in "support for traditionally 'female' roles like nurturing, supporting and cooperating ... to make computers serve humans" (Waltz, 2005, 96) rather than the reverse.

Marleen Stikker in the Netherlands was also interested in aligning cultural frameworks with hacktivist tactics, and so with the blessing of Amsterdam city officials in 1994, most of the country's best hackers went to work for her in building *De Digital Stad, the digital city*. There was much interest at that time in trying to create new kinds of translocal urban space, but Stikker's was unique in that it wasn't built for computer nerds, but for everyone. Between 30,000 and 40,000 people registered right off and applied for email accounts (this was when email was *au courant*). Users could create their own 'squares' (public spaces) or 'houses' (semi-private spaces). They could also meet other people in the MOO's interactive coffee houses to collaborate, play, or just converse. By the year 2000, the system had over 160,000 subscribers. Being similar to Second Life (minus the shopping):

> the "Digitale Stad" operates largely on the basis of the <city metaphor>；. The <content> on the DDS server borrows from the evolved structure of the city of Amsterdam: there are <pleins> (squares) and <digital homes>, chat functions are called <cafés>, there is a special <cemetery> for people who have left the Stad and the police have even opened their own … station. (Media Art Net: Digitale Stad)

The simulated City's ability to function as a political entity was tested when the Church of Scientology charged them with copyright violation. The citizens held a town meeting, launched protests, demonstrations, and printed T-Shirts, placards, and wallposters with sexist and embarrassing scientology teachings to lobby the government. In the real world, the Church was "forced to back off" (Waltz, 2005, 99). From the first, the City flourished because of how it embraced multiplicity in points of view. Where the Dutch government was cracking down on social protest and activism in day-to-day politics, it left the City alone. After its peak in 2000, it was gradually taken over by consumer culture forces and by IT consultants, and the old community slowly moved to other venues like Second Life, which launched in 2003.

Another grass-roots electronic media group that "wired up women's groups in the 1990s" was Barbara Ann O'Leary's *Virtual Sisterhood*. The goal of O'Leary's NGO Women's Environmental and Development Organization (based in Manitoba) was to foster grass-roots activism on the UN agenda in time for the Beijing UN World Conference on Women in 1995. At the time, grass-roots organizations used fax campaigns to get their word out. Faxes in less fortunate parts of the world, however, were very expensive and so Latin American members advocated for North American members to get with the program and get email (Waltz, 2005, 100). O'Leary was able to set up networks on the ground in Beijing (pretty amazing in a post-Tiennamen Square massacres city) to offset Chinese censorship and to allow the conversations to continue without respect for borders after the conference had ended (Waltz, 2005, 100). The Chinese government refused to allow new accounts to be created during the conference and so Chinese "organizations relied on borrowed accounts for Internet access" (Siphon did similar things during the Twitter revolution in Iran out of Ottawa; Waltz, 2005, 100). O'Leary and her people made sure that the daily information feed reached her Linkages server space back in Canada (which was largely donated by sympathetic, politically active corporate women), and there the information could be disseminated globally. Talk became action as O'Leary empowered a global women's movement. Her early implementation of RealAudio allowed them to put recorded interviews online. It was just-in-time kind of news. The Chinese government started to get nervous and cracked down on suspected dissidents and NGOs by moving the NGO part of the conference to a small town down the road. This made the electronic network even more essential. Ten years later her organizational structure was still seen as being radical. The surprise for O'Leary was the discovery of how innovative women organizers and dissidents were in the face of oppression: from shortwave radio in Costa Rica to fax campaigns which ran under the government radar in Beijing. In Eastern Europe they hooked up with Z-Net and reached places where direct contact by phone was illegal or impossible. After the conference, she continued to use hybrid technologies that merged fax, radio, and mail networks together with the rising electronic ones. The cyberfeminist troupe SubRosa has done similar things in the Third World for years.

All of these women worked to expand the "possibilities for activism online" by reinventing and reshaping online environments. O'Leary calls for us to, "'Maximize the internet's potential for women who are organizing for social change'" (Waltz, 2005, 102). Hacktivism is a term that was first coined in 1998 to describe an emerging hybrid form that united the best attributes of peaceful social protest – activism – and tech-savvy online civil disobedience – hackerism. It is a solution-oriented form of political action that inserts bodies and media-based dissent into real-time material concerns. It should not be confused with its adolescent and illegal cousins cracktivism – (code cracking, vandalism, data blockades (DDos), and the loss of digital data) – or cyberterrorism – (acts and agents of wanton destruction including worms and viruses). One of its trademark features is that the Web cannot contain hacktivism's flows, allowing it to spill out into the world in the form of political protest at WTO and G8 events, for example, and in books, pamphlets, net.art, and performance art. But hacktivism is only able to create disturbances and not real political change (lest this become a new Spectacular culture). There is the real importance in connectivity and social networking in our flesh-and-blood lives. Feminists realized this a long time ago.

Blogs are the predecessors of social networking sites like Facebook and Twitter, and have come to replace the bulletin board, the personal home page, the listserv, and email in turn as primary sites of connectivity. Social media are not simply soapboxes, but are places where users are participants and participants are users. YouTube, for instance, has fostered a new genre of social video. The problem with many 2.0 sites, like Facebook, is that they are instantiated in their medium. Take Second Life, for instance, which has its own currency, designer boutiques, and where virtually everything can be bought and sold and which can become the site for the creation of a new vision of cities as a countercultural space in Chinese artist Cao Fei's hands, as I will discuss in Part Three. Social media platforms are becoming part of a walled Web. They have donned their medium-like cement galoshes and do not usually look outside of themselves to critique the nature of their form. Too often social media and their creators fall victim to burnout or they are lured away to the next big corporate thing. The thoughtful critique and commentary gradually erodes in time to be replaced by lists and cut-and-paste content. Unlike Debord's ideal of cheap art, this is just serial linking. It does foster conversation, but in a time of information overload it does not necessarily foster reflection. Enabled by content management software that has removed the need for the average user to engage with code on the Web, many users of social media are guilty of not taking their message far enough. Debord's theories show how revolution is not possible in the printed medium alone, even if it is outside of commodified culture and the economy of exchange. That is to say, words alone are not enough. What they require are media tactics.

Tactics are not strategies, since strategies are goal-oriented. Tactics are fluid and adaptable disturbances, impolite ways of being in the world. Where blogs blend the old vestiges of print and oral telling to reinvent them as just-in-time commentary, and stick them on a monthly archival pin for a personal network to admire, tactical media instead use reverse engineering, open access, collaboration, and hacktivist approaches to disturb. Tactical media "are pliable and that pliability allows for on-the-fly critical intervention: statements, performances, and actions that must continually be altered in response to their object, 'constantly reconfigured to meet social demands'" (Raley, 2009, 6). Tactical media use peer2peer methods often in conjunction with 2.0 or mobile technologies to attack or critique corporate or political power. They may also provide alternative conduits for information flow or its backwash.

Social media are neither a form of activism nor a substitute for activism. As a social environment, they are tools that can be brought into play as a form of critique in tandem with activist tactics, but, as Geert Lovink puts it, "social movements do not emerge out of the Web. Their beginnings lay somewhere else, not in the act of online communication" (2008, 218). While they can make the personal political until they are blue in the face, disturbances must combine action and event to continue to speak louder than words. And activism itself is dead, or so Michael Hardt and Antonio Negri would have us believe. They say in their book *Multitude* that in our times "basic traditional models of political activism, class struggle and revolutionary organization have become outmoded and useless" (Hardt and Negri, 2004, 68). Soapbox 2.0 poses just this kind of problem if it is unresponsive and our only media tactic. Its practitioners are in danger of remaining on the sidelines of the actual, becoming virtual critics who preach to the converted if they use 2.0 methods only for propping up their ego. They are free to exercise freedom of speech, but without the equal and opposite power to realize the freedom of their ideas, as Geert Lovink discusses in *Zero Comments*. So if old school activism is dead, what choice do we have but to reinvent new tactics at the same time as we are reinventing ourselves? Part of activism's power was that it was site-specific. It identified a particular problem at a particular time. Clay Shirky has observed that "today's media artists often create software tools that are geographically specific. He calls such tools *situated software*. The anywhere and nowhere of the Internet is challenged by site-specific software art that addresses a particular community or location" (Lovink, 2008, 221). Hacktivists understand this better than just about anybody.

Hacktivism as a praxis was born in December 1997 when Critical Art Ensemble member and software engineer Carmin Karasic was so appalled by the events of the Acteal Massacre – where 45 Zapatistas were murdered at the hands of the Mexican government – that she set out to create a Web interface that would perform political protest as an aesthetic act. Together with three other Critical Art Ensemble members they formed the Electronic Disturbance Theatre. Their electronic civil disobedience engine is named FloodNet. Filling the browser page with the names of the dead, this activism tool "would access the page for Mexico's President Zedillo seeking bogus addresses, so the browser would return messages like 'human_rights not found on this server'" (Cassell, 2000). Unlike the attacks launched by crack-tivists, no damage is done by this software agent. When the Electronic Disturbance Theatre alerts its "online activists to 'commence flooding!'" they visit EDT's website and click on FloodNet's icon (Harmon). The software then directs their browser to the target, and cues the same page to load over and over again. In 2009, this engine was reinvented and turned on to Iran, and for eight days from June 13 until June 21 was targeting Iranian government sites.

As a tactical media disturbance, it is no accident that FloodNet must function as a community-based performance:

> FloodNet's action only drew its validity from the number of people showing support. "It was only actualized through thousands and thousands of participants," she remembers. "It was meaningless without the masses." Popular support transforms a random act of vandalism into a show of presence, Karasic argues. "This is an important difference between the single hacker/hacktivist who takes down a server with a single script" (quoted in Cassell, 2000).

Karasic sees her collectivity interface as something more closely akin to "conceptual art" than to cyberterrorism (Harmon, 1998). As Christine Paul has pointed out, digital media art has deep roots in conceptual forms.

Tactical media do not just report events; they are always participatory (Raley, 2009, 23). Power and protest have become unhinged in time and space: "it is fluid, decentralized, capable of resituating itself" (Raley, 2009, 44). The powerful places, architectural monuments, that housed power of old are defunct, and so too are the streets, which Rita Raley calls "dead capital" (Raley, 2009, 44). In fact, not only is activism outmoded but public space itself lies in ruins (Raley, 2009, 44). Protesting in the streets will not stop, as we saw on Wall Street in September and October 2011, in Egypt in January 2011 (which brought down President Mubarak in 18 days), and in Iran in 2009. Instead we should be waiting for a new wind. We should be waiting "to see a proliferation of hybridized actions that involve a multiplicity of tactics, combining actions on the street and actions in cyberspace" with a new species of social networks and modes of public disobedience (Raley, 2009, 44). Just as the hardware era is ending to make space for a wireless cloud, so the new flow will excite friction as it gathers speed. That 'new wind' may well be blowing from Asia and from South America, arising as it does out of different models of creative practice and new modes of translation.

Capture/leakage (performance + documentation)

In 1963, Naim June Paik stuck 50 strips of reel-to-reel audio tape to a wall, wired them up to speakers, and gave gallery goers a playback head so that they could make random music. That installation piece was called *Random Access.* According to Christiane Paul, "American digital artist Grahame Weinbren has stated that 'the digital revolution is a revolution of random access' – a revolution based on the possibilities of instant access to media elements that can be reshuffled in seemingly infinite combinations" (Paul, 2008, 15). Paik's piece is, and is about random access. It is also pure documentation – a subject that I will return to shortly. Random access is, of course, the *raison d'être* for sampling. Random access is the creative mode that makes the piecework, the reorganization of bits that is sampling, possible. Sampling itself is a live performance activity that is performed more and more in the moment in clubs like its progenitor turntablism. Random access is a shift from time-based forms of media (e.g., like film) to the spatialized or digital forms in the new media. It is also a shift from live to canned or mediated art. Prior to the media revolution, we lived lives of wholly unmediated access. Our interface with the world was our selves and our bodies. Everything was live. First mass media and then digital technologies changed that so that now our experience of everything is mediated. The resistance in digital tech is in finding ways to open holes in the mediated experience so that we can experience, if not live events, at least the semblance of live in liveness.

Liveness is about real-time intimacy, about shared experience and the notion of unrepeatability. This is not just a method of escaping the copyright cops, but a means of enacting disturbances in the corporate controlled media. Live performance, theorist Peggy Phelan says, honors the notion that an experience can be shared in real time that leaves no visible trace after the fact (Phelan, 1993, 127). For her, in theatrical performance, absolute repetition is simply not possible and as a result is a fraught concept (Phelan, 1993, 127). Representation can never perfectly reproduce the real, she argues; there is always a gap between them. In our time, it is precisely this *inability* of performance art to be re-created or archived that has made it a preferred medium for countercultural artists – like McLuhan and Debord – who sought slant strategies through direct interaction with a live audience (Auslander, 1999, 42). Only the documentation of the event survives. The Situationist International's slogans and McLuhan's probes are the vehicles through which an uncorruptable and nonrepresentational image was launched.

Liveness was, of course, born of the archivability of recording technologies. Without canned material, the notion of the "not live" makes no sense (Auslander, 1999, 51). Restrictive copyright regulations have now moved the art of sampling for the majority of composers (after the decline of the original hip-hop style due to aggressive policing of copyright) into live venues where the artist has become composer, VJ and DJ, and performer all at the same time. The VJ is a new kind of renaissance man in a swirling, multi-perspectival world. Where theater and traditional forms of performance used to be founded on a separation between the audience and the performer, more and more the contemporary form of performance art seeks not only to collapse that distance, but to draw in the user as an 'authentic', embodied interactor who engages with the performer either through cheap DIY media, dance in the club, or across the 'gap' of links or montage, or through other kinds of interaction in a gallery setting. Girl Talk's live performances are trademark liveness experiences. Gillis assembles sound loops in advance and then remixes them live in a club setting as a dance experience. Performance art seeks to transform the environment – the very walls of the medium's container – into something responsive to its audience. Like McLuhan's break boundary sound waves, "Being is performed (and made temporarily visible) in that suspended in-between" (Phelan, 1993, 167).

"New media theorist Paul Virilio … suggests in his book *The Vision Machine* (1994) that the use of surveillance technology by artists has resulted in the 'complete evaporation of visual subjectivity'" (Rush, 2007, 233). We can see this play out in a work like Jim Campbell's *Library* (2003). No stranger to creative remixing, Campbell is the creator of works called *Portrait of A Portrait of Harry Nyquist* (2000) and *Portrait of A Portrait of Claude Shannon* (2001); the two works are based on Nyquist and Shannon's "algorithm that describe these

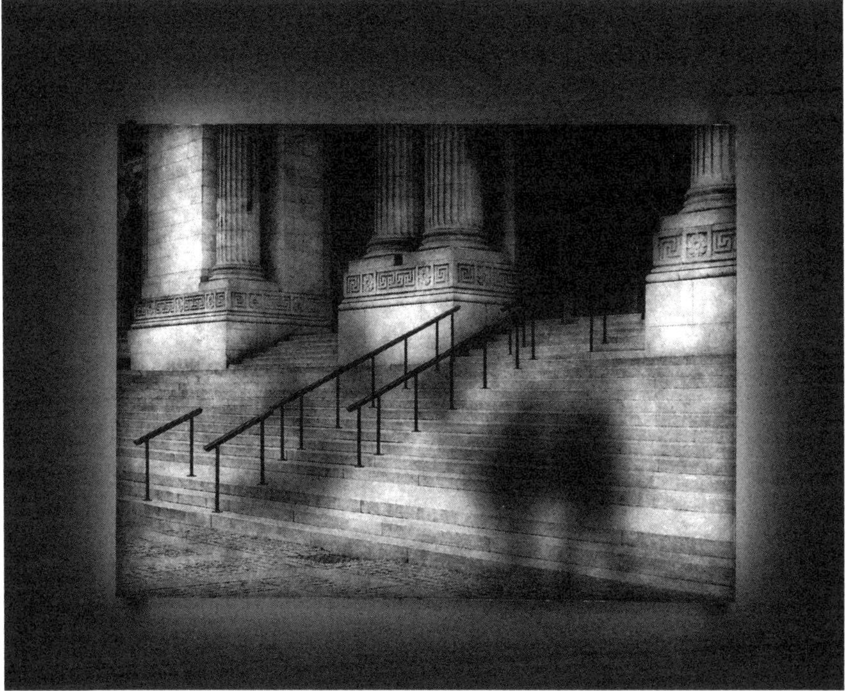

Figure 2 Jim Campbell, *The Library* (Photo credit: Sarah Christianson)

liminal transformations between analog and digital, between perception and illegibility" (quoted in *e-Art*). The portraits cycle through random noise, moving from barely perceptible to slightly perceptible. *Library* is a beautiful work that is a study in surveillance, motion, perception, memory, and loss. An image of the front steps of the New York Public Library is printed at a high resolution on rice paper, and suspended in a plexiglass frame that hovers in front of an LED monitor playing a low resolution 25-minute looped video sequence. The images are derived from surveillance cameras, and the rice paper cover renders the figures (humans and birds) into ghostly images that haunt the steps or flap in and out of the image. A psychogeographic meditation on presence and the presence of data trails in the city, this work brings home the fact that we never know when we are seen, tracked, or captured. Or who is watching. Where is the self in such a work? What or whose subjectivity is being enacted here? Who is performing? Is it the camera or the captured subject/object? Who is leaking the data? Whose subjectivity are we performing or inhabiting as we, as spectators, in turn don the proffered surveillance lens? What kind of information network is born out of these connections?

Live performance creates a community and connections where the body is highlighted as the sensory interface for our intimate conversations with the public and the private. This is a manifestation of the body as visible break boundary to be sure. Live performance art that uses the body as a medium is an expression of the writings of presence and of liveness in psychogeographic space. They undercut the distancing effects of technology through their immediacy in multi-mediated environments. Where even the live is mediated in our high-tech world, the only true or authentic experience is to be found in the body reimagined as a textual object or in the act of DIY social media. Performance art itself arose out of a particular moment in time with a political context. Amelia Jones explores the notion of the body as a performance medium for exploring subjectivity in her seminal work, *Body Art: Performing the Subject.* In these artists' hands, subjectivity became: "Embodied rather than transcendental, as in process, as engaged with and contingent on others in the world, and as multiply identified rather that reduced to a single, 'universal' image of the self" (1998, 197). We will see how differently this 'universal' plays out in the work of Chinese artists in Part Three. Performance art was a preferred interventionist practice for post-war revolutionaries, for instance, because it "intersects as well with the revolutionary challenges to patriarchal culture and cold war ideologies of individualism by the civil, new left, students', women's and gay/lesbian rights movments from the late 1950s onward" (Jones, 1998, 197).

McLuhan too understood that the body, as an extension of both our mind and senses, is the original medium – our sensory interface with the world. In a networked age more than ever, the role of the body as a medium for art and self-expression is integral to everything that makes us human. The body was both the first technology and the first medium of expression. Christian myth and ritualized practices have long been important vehicles for uniting transcendent body and mind in symbolic acts that are extra-linguistic. Performance *art*, however, takes this age-old tradition and situates it at a greater remove from these traditional acts by exteriorizing the internal psychic experience and replaying it as a public event. For McLuhan, the content of one medium was always another medium, and both old and new containers are transformed by their intermingling as an embodied performance. The real content of any medium, he advised us, is its *user*. Embodied, we are always already the content of our media, performing as witness our engagements with our linguistic and physical limits. As Susan Sontag noted, "The basic unit of contemporary art is not the idea, but the extension of sensations" (Sontag, 2001, 300). With film, for instance, the ultimate weapon of spectacular culture, its aesthetic is pure sensation, affecting us directly in our perception of time, movement, and memory. Because, Sontag says, cinema has more than content, it also becomes an ideal foundation for a theory of embodiment. The language of film is composed of camera angles, frame compositions, montages, cuts, and the remixing of familiar ideas or even found footage. The performance artist instead makes sensation visible. It is something that we witness as an externalized theatrical event. Collapsing the space between public and private, interior experience becomes a shared, public experience witnessed live.

At the moment at which Jackson Pollock splattered rather than brushed paint onto the canvas, the most important art form for the late Twentieth and upcoming Twenty-first century was born. Pollock's performative gestures known as "action painting" brought together the ancient tradition of painting with the act of the creation of the work of art. In the 1960s, performance art was established as a "distinct and definable" medium by Judy Chicago and Miriam Shapiro (Sayre, 1989, xiv). They experimented with the idea of performance during the 1960s, and by 1970 stunned the art world with their conceptual art installation *Womanhouse*, a navigable space with each room simultaneously architectural, sculptural, and interactive. Performance art is favored particularly by countercultural voices because it is ideally situated to undertake oppositional speech since each performance constitutes an unrepeatable, unarchivable, unique event. Witness, for instance, Sam Taylor-Wood's re-creation of Pollock's action painting in her photograph called "A Gesture Toward Action Painting" (1992).[1] One of the questions that Taylor-Wood's photograph raises is: Is this documentation or a finished product? Is this a performance or an art object in its own right? Ursula Frohne says,

> In Pollock's gesture and its replication, a number of categories circulate, such as those of the theatrical, the notion of performance, *mise-en-scène*, repetition and the transformation of the 'the image' (the screen) into a theatrical space or *mise-en-abyme*, as well as the transformation of the observer's perspective to a range of experiences made up of differing (art) historical perspectives. [...] Pollock's drip technique – an act of painting liberated from the control of reason – eliminated the boundaries of the picture's surface (Frohne, 2008, 355).

That is true enough for Pollock, but what about Taylor-Wood? Where Pollock's pieces were deemed masterpieces, the pieces in this photograph do not even look like Pollock paintings. In her C-Print, the 'painting' is virtually indistinguishable from wallpaper. Is this a commentary by Taylor-Wood on domestic space? A feminist performance of gender identity? Gender bending, gender confusion or gender appropriation? The artist's androgynous name and appearance as motion-blurred genderless 'Pollock' only underlines the ambiguity. Taylor-Wood becomes Pollock, but as a blurred subject, a dematerialized subject. Dare I say a virtual subject? According to Pollock, "the artist literally stepped into the picture" (Frohne, 2008, 356), but what kind of counterfeit is being wrought when Taylor-Wood performs Pollock in the act of breaking the frame of the painting *via photography as documentation*? Unlike Pollock's product, Taylor-Wood's process is lost and only the post-conceptual documentation of the performance of the act of breaking the frame of the work and of gender remains. Performance comes to play a different role here in the same way that the medium of photography does in relation to painting. The visual arts were such a revered art for so many centuries because the creation of an image, of a likeness, was so difficult. Now anyone can capture a moment, a frame, an object, or an event at any time. The making of an image is, therefore, no longer the difficult or unattainable part of an art form. The conceptual frame is now the difficult part and in an age of electronic disembodiment performance is foregrounded as a primary means of expression.

The artistry and the special nature of performance lies in its ephemeral nature combined with its ability to capture a process in motion. Its very liveness rejects the nature of art as something to be preserved, archived, and collected. It rejects the history of art as the creation of objects, and it rejects the self-defined role of the art establishment as something that commodifies *objets d'art*. Taylor-Wood's motion-blurred 'picture' of her performance of action painting and being-Pollock underscores the loss of authority in art and demonstrates how archiving is irrelevant where performance is king. With the same performative gesture, performance art uses its original, one-time nature instantiated in an embodied subject as a road back to the *raison d'être* for art.

But what happens to performance in digital space? What is liveness when it is entirely mediated by virtuality? I have already discussed Eva and Franco Mattes' fraudulent acts and hacks earlier in this chapter, but it is important to consider their works in the context of 'performance' as they are intended to be read. Some of their most celebrated works have been carried out in Second Life, the online world "where body, space and time can be completely reinvented" (as they say in their artist statement for "7,000 Oaks" (*Rhizome*)). They call them "Synthetic Performances" and they started in 2007. As with their Catt piece, "The 13 Most Beautiful Avatars" and "7,000 Oaks" restage or remediate works from other times, contexts, artists, and media. "The 13 Most Beautiful Avatars" is based on a film by Andy Warhol. Renowned for his so-called "Screen Tests" or filmic portraits, Warhol shot a lot of film footage between 1964 and 1966 of young aspiring actors, artists, and groupies who hung out at his Factory. Some of them ended up in his film *The 13 Most Beautiful Boys*. That title itself (it has been speculated) was probably derived from a New York City Police Department brochure, "The Thirteen Most Wanted" – which was itself the inspiration for Warhol's 20' x 20' silkscreened mural, *The Thirteen Most Wanted Men*, at the New York Pavilion at the 1964 World's Fair in Queens. So already their project is a remix of a work with a genealogy of versionings. The Mattes hung out in Second Life for a year and photographed the most beautiful avatars whom they encountered: these are the finalists (exhibited on November 15, 2006 at Second Life's Ars Virtua Gallery & The Italian Academy, Columbia University). The whole process of assigning people's own avatar creations a ranking on the basis of the off-world values for 'beauty' merely underlines how synthetic and entirely artificial these beauties are. For instance, there are no human aberrations, dogs, cats, or other SL creatures who make the finals, although one particularly angelic one has wings. The most beautiful avatars are as beautiful as beautiful people in the real world, which only serves to underline how artificial they are. The Mattes are very aware of the politics of avatars and have used them for other experiments. At their website, they state that these portraits of avatars address notions of identity, and assume:

> [T]he new online environments as places to socialize, nurture celebrity, and perhaps leave one's real self behind, these images capture members of the popular online virtual world *Second Life*, combining the traditions of glamor photography with the brilliant colors and hard-line aesthetics of the game-world. The Mattes' work questions both the traditional role of portraiture and the nature of the morphing relationship between identity and public presentation in virtual worlds. Like Andy Warhol's legendary Factory, *Second Life* is about the creation of alternate identities, of building and living a fantasy (Mattes website).

They also realize Warhol's dream of using really big printers, an aspiration they quote at their website: "'If printers ever get really big, like a twenty by thirty or thirty by forty, then it would really be great'. ~ Andy Warhol."

As members of Second Front, a Second Life-based performance troupe, the Mattes (or 0100101110101101.ORG as they are collectively known) staged a bank robbery at the Linden Treasury (Lindens are the currency of Second Life) in 2008. It was called *Grand Theft Avatar* and in that performance the group took on the semblances of a variety of new media artists and theorists, including Camille Utterback, Char Davies, Howard Rheingold and Christiane Paul, as they remixed the iconic video game.[2] Was this the first incident in Second Life of identity theft? They ended the performance by simulating the ending of Kubrick's 1964 film *Dr Strangelove, Or How I Learned to Stop Worrying and Love the Bomb,* by riding a bomb, *a là* Slim Pickins, into oblivion. (A favourite subject for remixing; see also Soda_Jerk's re-creation of *Dr Strangelove* in *HILTSWALTI: How I Learned to Stop Worrying and Love the Internet,* and Gustaf Mantel's animated GIFs out of Kubrick's most recognizable scenes (Lanks, 2011)).

As with all of the Mattes' work, the pieces tend to foreground ambivalence and 'raise as many questions as they resolve,' as new media theorist Patrick Lichty frames it. He says,

> The creation of remediated, "synthetic" works (to quote the Mattes) follows a strong contemporary tradition of reiteration of performance-based works in order to preserve their degree of affect in space and time. However, the remediation of performance/time-installations in virtual spaces tends to raise as many questions as they resolve. It straddles the formal grounds of avatar as body proxy and the viscerality/immediacy of performance and installation art (Lichty, 2007).

The Mattes call their Synthetic Performances in Second Life "online live gaming sessions" – even though Second Life is not a game. As I demonstrated with their mock online suicide, their play can be deadly serious.

Another of their works that I have already mentioned re-created Joseph Beuys' "7000 oaks," which was a massive tree-planting project begun at Kassel at Documenta 7 on March 16, 1982. Each tree was paired with a column-shaped stone monolith made of basalt. This was a massive environmental project and Beuys had intended to keep extending it around the globe "to effect environmental and social change" (Press release, Rhizome). As a re-enactment, the Mattes planted the first virtual tree with its stone companion exactly 25 years to the day (in 2007) after Beuys' first tree was set in the ground. In Second Life, the Mattes have their own island called Cosmos Island and all 7000 stones are there to mark the progress of the project as the pile diminishes. This is a participatory project and the trees that have been planted are responsive, and by touching them you can find out whose avatar planted it and when. As an environmental project, of course, it is ironic in the extreme. Lichty (2007) says,

> The key irony of this work is the cultural function that Beuys' 7000 Oaks has versus the Matte's remediation. While the form is recreated, and perhaps creates a method of awareness of environmental issues through Second Life, but according to calculations by Julian Bleecker, each avatar generates a carbon footprint equal to that of driving an SUV 1293 miles.

Irony is the Mattes' stock in trade as is so often the case with Twenty-first-century remixers and restagers. Consider, for instance, Keith Townsend Obadike, who used eBay to sell his 'blackness' in September 2011[3] before the site put a stop to the auction, or Keith Sanborn's work, *The Artwork in the Age of Mechanical Reproducibility by Walter Benjamin as Told to Keith Sanborn* ©

1936 Jayne Austen (1996). Sanborn's work is "made up entirely of the warnings about copyright violations that precede movies on video and DVD" (Yeo, 2004, 25). Unlike social commentators like Obadike or found-footage toilers like Sanborn though, the Mattes' work is not politically incendiary or "illegal." It is just not *original* in the old sense of the word. It is, however, highly inventive, thought-provoking, and very amusing. As conceptual art, the idea is king.

In a series called *Paradise Ahead* (2006–7), Scott Kildall, another recombinant performance artist, has set himself the task of re-creating all of the great performance art pieces in Second Life. His series includes remediations of: "Claim" (Acconci); "Leap into the Void" (Klein); "Shoot" (Burden, done with Jeremy Owen Turner); "Fall II" (Bas Jan Ader); "Cut Piece" (Yoko Ono, done with Jeremy Owen Turner); "Lift" (Fiona Tan); "Electric Earth" (Doug Aitken); "The Ninth Hour" (Maurizio Cattelan); "Rest Energy" (Marina Abramovic & Ulay, done with Jeremy Owen Turner); "I Like America and America Likes Me" (Joseph Beuys); "The Staircase" (Peter Land); and "Remake" (Pierre Huyghe, done with Jeremy Owen Turner). Like Taylor-Wood's Pollock and the Mattes' *The 13 Most Beautiful Avatars,* this is a print series that documents the performance after it has finished. When the work becomes pure simulation of a restaged live, embodied event, what is left? What is the meaning? Are there any bodies involved? Are we merely held transfixed? Lichty (2007) says,

> The works become ambivalent, playing with our affective connection to the avatar while not actually sinking the blade through the skin, shooting the bullet, etc. It holds us in the anxious moment before the trigger pulls, the razor cuts, with the historical referent and the gesture all in place, with our memories of the vestigial flesh and its history holding the gut taut.

What Kildall sets out to do, he tells us at his website, is to blur "the line between document and event." Unlike the Mattes who actually stage *performances*, Kildall stages events for *documentation.* "Using only [the] primitive graphics of Second Life, the documentation of these performances," which exist only as large-size prints, he says, "serve as a historical record of the initial launch point into simulated worlds." And while they are re-creations of the original performances, Kildall's avatar is about 7 feet tall with pink hair and purple skin, so there is no attempt to re-create the 'look' of the original event – only to document its resituation. What is the role of embodiment in the documented event? Patrick Lichty (2007) says,

> [I]t seems to me that separating the issues of embodiment and context are problematic at best. It seems that in the case of virtual worlds like Second

Life, that the ironies created by the remediation of works is closely linked to the juxtaposition of the relationship between the physical and virtual. For example, if Abramovic's Red Star were to be recreated through artful retexturing of the avatar and proper posing, what then becomes of the investiture of the flesh? One could say that the gesture becomes wholly symbolic, but residents in Second Life clearly have investiture in the avatar as extensions of themselves.

Second Life is a space for the exploration of identity politics, and, in a virtual world, one's appearance, one's virtual body, is one's representation and the documentation of an existence. More than this, part of the pleasure of these kinds of documented events becomes the act of doing it oneself. DIY – which includes the analog – is the new preferred aesthetic.

Critical Art Ensemble documents the resurfacing of the analogic as a mode of art creation in a new DIY economy. It is the luxury economy that for a long time has preserved the semblance of the handmade: "This is the area where one-of-a-kind, customized, and designer products still rule over the cheap imitations and digital knock-offs" (Critical Art Ensemble, 2000, 153).

But everything old is new again. In 2011, SIGGRAPH's art show, curated by Andrea Zapp, was called "Analogue is the New Digital: An Online Exhibition." The digital DIY now threatens to perform a reversal on those traditional values:

[T]he anachronistic economy of artisans reproduces itself as luxury economy. [...] Custom-made jewelry, haute couture, and high art are still the signifiers of privilege that underlie aesthetic value. They are perfection in a world of counterfeits. The luxury market is closely related to high culture, but as we shall see, this secular field of sacred privilege is also being quietly plundered by the digital (Critical Art Ensemble, 2000, 153).

For the digital DIY means subverting the spectacle and providing a means of producing new media, media-as-process that exist only in their documentation – like the #occupywallstreet event that the mainstream media find unnewsworthy, unnewsworthy because it subverts the gaze, because it is all documentation. It was this impulse that also gave birth to tactical media. Tactical media were born at the point when reality and representation collapsed together. Matthew Fuller in "Towards an Evil Media Studies" says that this is the rub. The politics of representation "are badly adapted to an understanding of the increasingly infrastructural nature of communications in a world of digital media" (Fuller, 2007) and yet representation is always already political. And this is where tactical media become a slippery slope:

> To believe that issues of representation are now irrelevant is to believe that the very real life chances of groups and individuals are not still crucially affected by the available images circulating in any given society. And the fact that we no longer see the mass media as the sole and centralized source of our self definitions might make these issues more slippery but that does not make them redundant (Garcia and Lovink, 1997).

Tactical media, as a result, are always participatory. This too is what de Certeau advocates in *The Practice of Everyday Life* (1984). Representations need to be replaced in our imagination by the uses of representations so that we can act "tactically." The mass media and popular culture are a shifting "set of practices or operations performed on textual or text like structures" (quoted in Garcia and Lovink, 1997). De Certeau resituates consumption as, I would argue, a kind of cannibalism, where the weak digest the strong. De Certeau's messy or leaky aesthetic joins the 'tactical' user – what we might think of as a kind of hacker or digital DIY proponent – with the intelligentsia or 'presumptuous producers' (including educators, writers, curators, and revolutionaries), who he dubbed "strategic," to produce activities that include "[p]oaching, tricking, reading, speaking, strolling, shopping, desiring. [And utilize] Clever tricks, the hunter's cunning, maneuvers, polymorphic situations, joyful discoveries, poetic as well as warlike" (Garcia and Lovink, 1997). It is in the get-down-and-get-dirty aesthetic that the new digital resistance is being spoken. David Garcia and Geert Lovink in "The ABC of Tactical Media" (1997) say,

> Tactical Media are what happens when the cheap 'do it yourself' media, made possible by the revolution in consumer electronics and expanded forms of distribution (from public access cable to the internet) are exploited by groups and individuals who feel aggrieved by or excluded from the wider culture. Tactical media do not just report events, as they are never impartial they always participate and it is this that more than anything separates them from mainstream media.

In the age of documentation, this is the realm of Michael Warner's third public, a public that comes into being around a cluster or collection of texts (Warner, 2005, 49, 50). The first kind of public is a form of "social totality" like a nation. The second kind is "a concrete audience, a crowd witnessing itself in visible space, as with a theatrical public" (49). The third kind is specifically bound to an event. Warner says that publics have their own rules of governance: it is "self-organized," "exist[ing:] by virtue of being addressed"; it is "a relation among strangers"; it is "constituted through mere attention";

it is "the social space created by the reflexive circulation of discourse"; it behaves "historically according to the temporality of their [texts'] circulation"; and it "is poetic world-making" (50–82). Counterpublics behave in the same way, except counterpublic discourse is more personal; those it "Addresses are socially marked by their participation in this kind of discourse; ordinary people are presumed to not want to be mistaken for the kind of person who would participate in this kind of talk or be present in this kind of scene" (86). A counterpublic is "a scene for developing oppositional interpretations of its members' identities, interests, and needs" (86). Counterpublics also feel themselves subordinate to the dominant culture (86). The DIY culture is the "kind of public that comes into being only in relation to texts and their circulation" (Warner, 2005, 65–6). Youth culture, like those participating in social media revolutions around the globe, and others taking oppositional stands as, for instance, in support of WikiLeaks are a new circulating and leaky body who increasingly define themselves in opposition to consumer culture in particular circumstances. (Consumer culture, of course, has no outside and so, like it or not, the allegiance always returns.)

Tactical media are slippery and, as Raley also notes, circulate around events that "focus on instructions rather than products" (Garcia and Lovink, 1997, 9). They leak and linger on in "traces and residues" (Critical Art Ensemble, 2001). WikiLeaks are the next stage in documentation, a global version of this viral practice of engaging in lowbrow creative and critical production. Soda_Jerk's *HILTSWALTI*, for example, explicitly explores "the way that discourses of war have been appropriated within the political and media arena to frame WikiLeaks as a form of information warfare" and "seeks to engineer connections between WikiLeaks and other historical vectors such as the Cold War, the Pentagon Papers and Watergate" (Soda_Jerk, 2011a). In a sense WikiLeaks (and resistance) had to go global because now protests (only) find their supporters online in the production of social relationships. WikiLeaks as a collection of or venue for counterpublics function like protocol does in the digital realm. Alexander Galloway in *Protocol* reminds us that networks are not metaphors, but "materialized and materializing media" that unite connectivity, collectivity and participation. An awareness of the material nature of the media is essential in practicing resistance against "power relationships in control societies" (Galloway, 2004, xv). In the net, protocol is an aesthetic force that has control over life itself. Protocol is always about disconnection as much as it is about connection. Disconnection as a force has been made all too clear in the move to unplug social media activists during recent events in Libya and in the Bay Area Rapid Transit system in California by surveillance societies. Protocol was also enacted as a way of shutting down WikiLeaks. There is a precedent for this kind of behavior in 2002, according to Galloway, who says this seems to have been used as a way of shutting down the culture jamming group The Yes Men via their ISP, thing.net or The Thing, a long-time nonprofit internet service provider for art organizations and the arts community, on December 4, 2002, The Thing were notified by their network provider, Verio, that they were in violation of their service contract. Many groups and projects went offline. Verio cited the cause as actions by the Yes Men's activities around the so-called Bhopal Incident in India; Union Carbide had a massive chemical spill which to this day has neither been adequately cleaned up nor the victims compensated. The Yes Men's violating activity was circulating a press release on the eighteenth anniversary of the disaster in which the parent company, Dow, disavowed any responsibility whatsoever. (Two years later on the twentieth anniversary the Yes Men would make an even bolder statement when one of their members impersonated a spokesman for Union Carbide live on the BBC news and took full responsibility for the event in India, which killed thousands and left more than 100,000 requiring lifetime medical care.) The press release was designed to start a conversation about the nature of globalization and global responsibility. Galloway says that the protocol under which thing.net was shut down – permanently – was the Digital Millennium Copyright Act (DMCA).[4] This protocol enabled a silencing from within the system's own operating procedure (Galloway, 2004, xvii). But protocol works both ways. It can also become a means of using the system to enact resistance. Capture is a protocol and means of control within capitalist culture whether it is done technologically, visually, statistically, or physically. But capture, as an immovable force, always has its opposite built in. As soon as capture is enacted, leakage is inevitable. It is just a matter of time. For instance, London (England) has more security cameras in place than anywhere else on Earth. They number in the tens of thousands and watchdog organizations estimate that there are in excess of four million cameras in use in the U.K. alone. Each London resident is on average captured on camera hundreds of times a day. In *CTheory*, John Armitage notes that Italian media theorist Paul Virilio says in his book, *The Lost Dimension:*

> [T]he rule in the overexposed city is the disappearance of aesthetics and whole dimensions into a militarized and cinematographic field of retinal persistence, interruption, and 'technological space-time'. Speaking … about the overexposed city within the context of the 'totally bogus' court cases surrounding O. J. Simpson and the death of Princess Diana, Virilio suggested that, today, "**all** cities are overexposed" (quoted in Armitage, 2002).

Surveillance technologies are mechanisms of power. Power mechanisms are in turn responsible for subject formation. While capture is something that we routinely do to data, it is also something that is done to us. Capture has lingering implications of surrender or entrapment or defeat (as in chess), but it is also used in reference to data: motion capture records movement; screen capture records images or processes depicted on a computer monitor; video capture converts analog video to digital form. Kriss Ravetto-Biagioli (2010, 124) says that

> Automated systems of surveillance – millions of CCTV cameras in bank machines, offices, schools, corporate and government buildings, at traffic lights, inside taxi cabs (in the United Kingdom) – do modify people's behavior. What it is hard to predict, however, is how to control, or even measure the effects of such modifications.

Within Spectacular culture, he or she who controls representation possesses the power to visualize data and shape his or her own narratives. Cameras capture data and produce low-res images and objects, which could be or could become aestheticized. What happens if we reverse the usual paradigm? What if surveillance technology is inverted and the subject takes back control of his or her own military-industrial data – data that are increasingly leaky? Refusing the spectacular model of cinema, real-time documentation, tracking, and surveillance are new digital modes of creation for Manu Luksch who seeks to track subjects as a means of creation. In her film *Faceless,* she creates a 55-minute film working with CCTV footage as 'legal readymades'. Using the U.K. Data Protection Act and various legislation out of the EU to her own ends, *Faceless* constructs a Data Protection Art. Following the protocol of her own "Manifesto for CCTV Filmmakers", Luksch uses these captured images, which anyone can request of themselves, to construct a surreal, choreographed world where everyone's face is obliterated (as required by the privacy law). In her "Manifesto" Luksch posits the guerrilla filmmaker as a viral agent and a "symbiont." She says that the

> most prolific documentarists are no longer to be found in film schools and TV stations. In some European and American cities, every street corner is under constant surveillance using recording closed-circuit TV (CCTV) cameras. Such cameras are typically operated by local government, police, private security firms, large corporations and small businesses, and private individuals, and may be automatic or controlled (zoomed and panned) from a remote control room. Filmmakers, and in particular documentarists of all flavours, should reflect on this constant gaze. Why

bring in additional cameras, when much private and public urban space is already covered from numerous angles? (http://www.ambienttv.net/content/?q=dpamanifesto)

In the act of hijacking these technologies, lighting and cameras become not just unnecessary but forbidden. The hack is complete once Big Brother's eye is used for aesthetic effects.[5]

McLuhan said that "technologies begin to perform the function of art in making us aware of the psychic and social consequences of technology" (1964, 14) and these artistic endeavors invert internalized psychic and social events to make them visible as art. As our technology more and more stands as an interface and a boundary object between us and a world where everything is mediated, ostensibly only live performance art can create disturbances that blend the act and event to speak our humanity as creatures of flesh and blood. As the body becomes an interface in our engagements with both art and technology, can the medium continue to be the message as McLuhan claimed? Convergence of media and machines is undermining old notions of media specificity in dramatic ways.

If we read McLuhan as a performance artist like Taylor-Wood, it is hard to interpret his performance art as anything but a break boundary. Break boundaries allow us to see the bones of the code beneath the surface of the new media's form, the rhythms in the structure. Was performance art McLuhan's method of finding ways of translating into visible form what is written on the outsides of himself, of ourselves (like social codes) into a performance that links what he and we tell ourselves inside (like identity narratives)? What is the meaning of inside and outside when all events including the physical are mediated? As the body becomes an interface in our engagements with both art and technology, indeed as we find ourselves so plugged in that we become a part of immersive environments, can the medium continue to be the message as McLuhan claimed?

In 1995, Nicholas Negroponte, Director of the MIT MediaLab and the inventor of the $100 Laptop, saw medium and message beginning to converge: "The medium is not the message in a digital world. It is an embodiment of it. A message might have several embodiments automatically derivable from the same data" (Negroponte, 1995, 71). In the digitial world those embodiments are documentation. Four years before that, in his seminal work *Simulacra and Simulation*, French theorist Jean Baudrillard too questioned the future of McLuhan's medium and message in our time:

[T]he medium is the message not only signifies the end of the message, but also *the end of the medium*. There are no more media in the literal

sense of the word (I'm speaking particularly of electronic mass media) –
that is, of a mediating power between one reality and another, between
one state of the real and another. Neither in content, nor in form. Strictly,
this is what implosion signifies. The absorption of one pole into another,
the short-circuiting between poles of every differential system of meaning,
the erasure of distinct terms and oppositions, including that of the medium
and of the real – thus the impossibility of any mediation, of any dialectical
intervention between the two or from one to the other (Baudrillaud, 1991,
82–3).

Performance and documentation are that dialectal intervention. As everything
becomes digital and modes of distribution change to reflect that, it is how
works are distributed rather than their medium that is coming to be a deciding
factor in how we use them. As Vilém Flusser states, *praxis* is transformed
by its channel of distribution, and not the reverse. (2000, 53). The public and
private threaten to disappear into televisual space just as the shifting roles
of the artist and the browser, and the work of art and its interactor call into
question the site of art itself.

Egyptian artist Khaled Hafez plays with just these parameters in his
5-minute, 30 second three-channel work, *The Video Diaries* (Mercosul
Biennale, Brazil, 2011). Situated in conjunction with a series of C-Prints, called
The Field Diaries, and a single channel video, called *Revolution, The Video
Diaries* exude revolutionary fervor inspired by the events that transpired in
Cairo from January to February 2011. Both of the diaries are concerned with
geopolitics and social media, but for Hafez these recent events are personal.
He "explores the themes of identity and nationhood in work that documents
the artists' own perceptions of moments lived during the country's 18 days
of revolution" (press release). The press release states:

The Video Diaries project combines video footage taken by Hafez with
stock footage extracted from social media and from several other sources.
These are assembled to create several parallel narratives that inter-
twine on the three screens as the real footage of collective doing, and
sometimes violence (*sic*). The flux of information disseminated by the
media footage, the lack of structured dialogue combined with real sounds
from the Tahrir Square, where the 2011 revolution takes place, all are
pasted with the sound of solo guitar music. Through this use of music,
the idea of "revolution" is romanticized, adding a simulated fictitious
atmosphere to the very real footage, to represent intimacy and personal
nostalgia" according to the artist. Hafez's choice of medium blurs the

distinction between the artist as detached observer and active participant in the events of revolution.

As we will see with cynical realism and political pop in Chinese revolutionary art in Part Three of this volume, the clash of public views and private politics is always complicated, and always involves a merging of public (exterior) and private (interior) space. Drawing on the 'aesthetic journalism' of Alfredo Cameroti and Khaled Ramadan, Hafez culls and remixes footage from television, surveillance cameras, satellite feeds, amateur video, and professional news sources with his own footage to construct a situation for himself to speak against the mix of official and unofficial narratives. His "stance reiterates Cramerotti's basic thesis that 'the blurring of margins between artistic and information practices is a main feature of contemporary culture'" (press release). An accomplished remixer and found-footage manipulator, Hafez has already explored mashing up levels of discourse and the commingling of highbrow and lowbrow images at length in his earlier paintings and multimedia works, especially in the two videos of *The A77A Project: On Presidents and Superheroes* in which he casts Anubis, the Egyptian dog-faced god, as Batman in an Arab landscape. The role of social media in the political events has been celebrated uncritically. Hafez urges us to look and look again at the visual rhetorical inherent in the environment that is the politics of revolution.

Hafez's work is not the only art that has emerged out of Tahrir Square. There has been a huge outpouring of creative innovation in software, organizing, and information distribution. The flood of this Arabian Spring has not stopped. Websites like *Mosireen*, *Take the Square*, and *Cinerevolution Now: Tahrir Cinema* are rich sites for the exploration of online activisms WikiLeaks style, and effusive expressions of creative digital energies. *Tahrir Cinema* is a Twenty-first century virtual public square where the organizers throw an online sheet up over the Web wall and offer screenings of events from the Revolution. Another fascinating innovation that came out of the social protest was *Stateless Plug-in*. Developed by Conradi and Christensen, it is a plug-in that runs on three browsers: Mozilla Firefox, Google Chrome and Apple Safari.

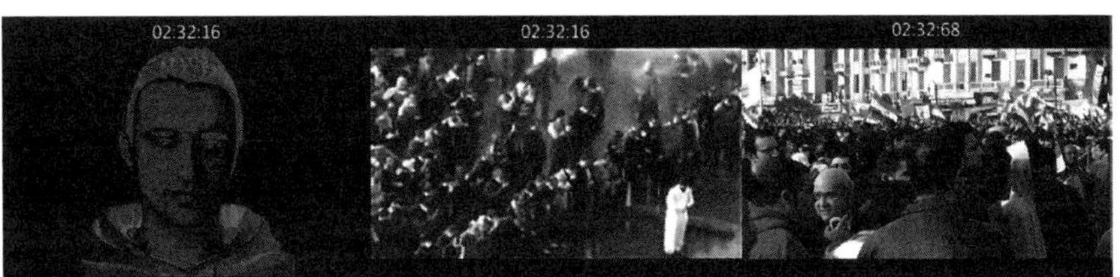

Figure 3 Khaled Hafez, *The Video Diaries* (2011)

Exploring the issues surrounding statelessness – people without a state to call their own, which total an estimated 12 to 15 million people in the world – and the resulting lack of rights, the plug-in territorializes your browser and highlights 30 words in digital space that are used to foreground the politics of this situation. The words (flagged in German, Dutch, and English) are space, mobile, sent, destination, access, arrival, departure, travel, directions, visit, options, profile, home, relationship, opportunities, security, search, privacy, safety, memory, terms, help, conditions, status, notified, register, report, categories, resolution, and contact. All of these words are transformed, Conradi and Christensen argue, when transposed or translated from digital space to the territory of statelessness. Together they come to revolve around issues of policy, mobility, and identity based on the cross-referencing of several sites, including Blogger, Flickr, the United Nations refugee Agency (UNHCR), YouTube, and the *New York Times*. Finally, the plug-in mingles highbrow and lowbrow forms of speaking so that all communication situations and environments are represented.

McLuhan said that environments were always elusive: they "are invisible. Their ground rules, pervasive structure and overall patterns elude easy perception" (McLuhan and Fiore, 2003, 84–5). As convergence has collapsed the senses into undifferentiated, clickable modalities, has it thereby rendered the senses obsolete? Or, as McLuhan maintains, has it simply created an inversion that foregrounds the user as content and the consciousness as the final frontier? Rosalind Krauss questions whether the term 'medium' continues to be of any use to art at all, reduced as it is since the advent of minimalism and with the rise of the digital where everything is always already virtual. Krauss ultimately decides that contemporary works of art exist as recursive structures in a 'post-medium condition,' and, since the nature of a medium is the sum of its "manifest physical properties," it ultimately, therefore, must "specify itself" or speak its own shape (Krauss, 2008, 7). Documentation and performance are the vehicles of transmission. Curator Peter Weibel says that what he calls "postmedia" art – or after-media – is what comes after a post-medium condition and the affirmation of the media themselves; with the effects of the digital now being universal, all media exist in an afterlife state (Weibel, 2006, 98). He states that the media have two stages: the first is a leveling and elevating one where all media achieve the same status; the second is the coming together, where they merge identities and intermingle in a symbiotic act:

This media experience has become the norm for all aesthetic experience. Hence in art there is no longer anything beyond the media. No-one can escape from the media. There is no longer any painting outside and beyond the media experience. There is no longer any sculpture outside and beyond the media experience. There is no longer any photography outside and beyond the media experience (Weibel, 2006, 98).

McLuhan argued that the media's next big thing would be the extension of consciousness. Has the medium collapsed these obsolete things called the senses into the message or has the sensorium been rewired with an added hyper-aware and self-reflexive dimension? It is so-called piracy, the reuse of familiar images to foster stoppage and repetition and critical thought alongside tactical media that stand as bulkwards against aestheticization, as depicted in the Wachowski brothers' film *The Matrix.*

Resistance is inherent in digital culture and its creative practice, according to the Critical Art Ensemble, a new media performance art interventionist and hacktivist troupe. In their article "Recombinant Theatre and Digital Resistance" they explore the digital as a method of enacting resistance. They say "there has always been an undercurrent of resistance in the digital," and that as early as 1870,

> Le Comte de Lautréamont published in *Poems*: "Plagiarism is necessary. Progress implies it. It embraces an author's phrase, makes use of his expressions, erases a false idea, and replaces it with the right idea." In three sentences Lautréamont summed up the methods and means of digital aesthetics as a process of copying a process that offers dominant culture minimal material for recuperation by recycling the same images, actions, and sounds into radical discourse. Over the past century, a long-standing tradition of digital cultural resistance has emerged that has used recombinant methods in the various forms of combines, sampling, pangender performance, bricolage, detournement, readymades, appro-priation, plagiarism, theatre of everyday life, constellations, and so on. Maintaining this historical tendency by further refining methods, finding new applications, furthering its theoretical articulation, and increasing its rate of manifestation is an ongoing task for those who hope to see the decline of authoritarian culture (Critical Art Ensemble, 2000, 151–2)

Digital cultural resistance for them is a way of shattering the spectacle and inserting the body back into the societal equation, as in Hafez's *Video Diaries.* It is the marriage of aesthetic practices and bodies that permit official power structures to be undermined and collective power to resurface. In remixing, original, truly creative voices emerge from unofficial spaces – the cultural borderlands, the liminal zones, the thresholds between cultures – and it is in those voices that they find possibilities for real change (2000, 157). The reason for this is that the value of digital resistance lies in the semiotic network that remixed objects create (2000, 155). Critical Art Ensemble uses Duchamp and his readymades as an example because the signed object is declared 'art' via the context of being exhibited – captured – and by the inherent network, to other objects and artists, it leaks through its situation. The signature is the documentation. This is the space of tactical media discussed earlier. The shock of seeing again in the readymade interrupts us and pierces the veil of the spectacle. The disturbance in its disruption of political power structures brings together the event and the politicized action, but it is in its performance, and the documentation thereof, that the process of capture and leakage can be realized. No "object has essential value" (2000, 155); it is only by capturing it and situating it in a particular context that we have a framework for appreciating and understanding art.

The implication implicit in this is that digital art is not just to be looked at or interacted with or performed. In its perpetual give-and-take, capture-and-leakage, it also looks back. Just as the commons has been foreclosed on by the privatization of property, so our public space (which includes social networking sites like Facebook) is increasingly controlled and policed by electric eyes and digital surveillance that capture us and hold us within their gaze. Mexican-Canadian artist Raphael Lozano-Hemmer's unique installations have explored issues of surveillance in depth. Surely none is creepier than his 1997 Venice Bienniale piece, *Surface Tension*. An interactive installation, the work is deceptively simple. It consists of a giant, all-too human eye that follows a viewer around the gallery space. Technically blind but creepier for its ability to blink, it is guided by an array of sensors that detect movement. It only stands down (and closes its lid) when viewers' backs are turned, escape detection, or are not present. It is relentless in the manner of Big Brother. That there is no intelligence behind it, only surveillance, in no way changes the dramatic effect it has on viewers who hide or run, or attempt to dodge the gaze. Another work by Lozano-Hemmer, *Under Scan*, reverses the parameters of the first project and through surveillance cameras endows an interactor in public space with a new shadow – the shadow of another person who appears to be sleeping. Drawn randomly from a database of multi-ethnic video clips that range from a few seconds to minutes, the computer calculates perspective and dimension to fit the image of their supplemented shadow to the user as he or she strolls through urban space. "*Under Scan* exemplifies what Paul Virilio calls the militarization of vision, which in turn militarizes urban space, reducing vision and urban space to a field of tactical operations" (Ravetto-Biagioli, 2010, 128). This process of capture is literally arresting. In "Shadowed by Images," Kriss Ravetto-Biagioli notes that

Lozano-Hemmer's work also responds to the fact that public space and public relations have already been territorialized by various commercial, political, and military regimes: "The urban environment no longer represents the citizens, it represents capital." Urbanism itself is just another mode of appropriating public space by capitalism's information economy, and a way of continually "refashioning space into its own setting" (Ravetto-Biagioli, 2010, 136).

While there is no outside in surveillance culture, the awareness of being watched raises awareness of the gaze itself and of the necessity of hacking one's way out. It is also generally accepted that government and the military are in collusion with corporations and it is common knowledge that technologies, tactics, and information are shared as a hard-and-fast rule (as Lawrence Lessig has explored at length in *Code: And Other Laws of Cyberspace* (1999)). Facebook has made ubiquitous the notion of social media as an information capture medium and most accept this unblinkingly as a part of their Twenty-first-century lives. What effects such a close capture and persistent gaze have on us is hard to measure, but with the arrival of WikiLeaks in 2006 a metaphor and mode of distribution have been realized that provide real alternatives for information dissemination.

Documentation is the art form of the Twenty-first century. One of the key questions in the role of conceptual art is always about the role and nature of documentation, as well as "the broader relationship of art to document, history, and photography" (Alberro, 2001, 5). Alexander Alberro notes that under conceptual art *art becomes information*. For conceptual artists, documentation is central to the work of art itself. If it is catalogued photographically is the photograph the work or merely capture, i.e. documentation? The whole purpose of an art catalog, for instance, is to reveal the nature of the work. If the gallery takes photographs for the catalog, where lies the work of art? Alberro says, "Catalogues … record ownership and property rights." As Weiner noted (in 1969 at Seth Siegelaub's show),

> anyone can copy anything that they want, exactly, so that "complete proof of receivership" is located only in the "title". Indeed the establishment of ownership is prominent, even conspicuous, in exhibition catalogues of the late 1960s and 1970s that feature Weiner's work: following the title and date of each work, the catalogues make a point of indicating the collection to which it belongs (Alberro, 2001, 6).

Is photography as an act of capturing or documenting a work of art *the* work of art? In the 1960s at the inception of conceptualism, the artists were very aware of the importance of documentation. Douglas Huebler states that "it's the documents that carry the idea" of the work (quoted in Alberro, 2001, 7). Sol LeWitt likened documentation to "a thought process" and "a musical score"; Robert Smithson called his documentation "a map" and "a trace" with any photographs becoming merely "an accessory to the event" (quoted in Alberro, 2001, 8). Alberro charts through these artists' views how even as early as the 1960s the notion of artwork and documentation were beginning to merge, that art was beginning to become information. While some of these early conceptualists, like Wiener, maintained that an "event can never be art" others were starting to sway in the other direction (Alberro, 2001, 9).

For us there is no such gap. Not only is everything information, but it is increasingly difficult to tell what is art, what is journalism, what is advertising, and what is surveillance. What nagged at the hearts of the early conceptualists was precisely this fear of leveling: "all cultural products function on the same plane, without depth, hierarchy, and intrinsic value or meaning" (Alberro, 2001, 12). Where the early conceptualists opened the door to interactivity, participatory culture involves a different, deeper, richer engagement with tactical media. McKenzie Wark, in "Strategies for Tactical Media," (2011) links the tactical with strategic and logistical media as well. He says, "Paul Virilio argues that in military affairs conflict has passed from dominance of the tactical to the strategic and on to the logistic." Now, he says, logistics controls everything from production to communication. Wars are strategies now waged on the basis of logistics by what he calls the "military entertainment complex." In the fact of such opposition, tactical media are essential for finding a way to hack the system and establish oppositional channels, or, in Geert Lovink's words, "temporary consensus zones." Lovink may be riffing on American writer Hakim Bey's concept of TAZ or a temporary autonomous zone that arises out of revolutions when there is a disruption in the hegemonic order. In the early pre-Web days of digital culture, Bey advocated poetic terrorism and art shock as a means of shattering the spectacle. In *Code: And Other Laws of Cyberspace,* Lawrence Lessig also warns that military space and vision are sites for the harvesting of all kinds of data, and that they share information with the government, the police, corporations, and whoever else might be interested.

More and more in data space, unmediated space is disappearing and militarized, and real space begins to merge. A poignant example of this is Josh Bricker's *Post Newtonianism (War Footage/Call of Duty 4 Modern Warfare).* The title is derived from a quote by Henry Kissinger, which Bricker says was one of the inspirations for his piece. Kissinger said in his essay "Domestic Structure and Foreign Policy" that it was the East's failure to embrace the theories of the Enlightenment, especially Newton's theories, that had led them to be so 'backward' and 'inferior'. Bricker combined this with his readings of Edward Said's *Orientalism* and Alan Lightman's *Reunion* to create his mashup. Bricker takes the WikiLeaks video of the American attack on unarmed civilians and journalists and combines it in two channels with the WikiLeaks footage on the left and a video stream from the game *Call of Duty 4: Modern Warfare* on the right. Both streams are seen as if through infrared goggles. The soundtrack blends 50 percent of the original, WikiLeaks soundtrack with 50 percent new or game-derived material. It is a chilling mashup that explores the desensitizing powers of war as seen through the mediated image. An exploration of the "media-driven simulacrum," Bricker says,

> My work is an attempt to expose the power structures that dominate our lives; those we witness, take for granted, and participate in (both consciously and subconsciously). I want to peel back and reveal traditional artifices, dominant cultural hierarchies, social systems and their effects, exposing their substructures like armatures in order to provide context for the creation of new, unexpected connections and personal truths (Bittanti).

For theorist Paul Virilio the possibility of a "'vision machine'" loomed "in the near future, a machine that would be capable not only of recognizing the contours of shapes, but also of completely interpreting the visual field, of staging a complex environment close-up or at a distance" (Virilio, 1994, 59). Once vision becomes sightless, cameras can be controlled remotely.

> Virilio's interests in war, cinema, and the logistics of perception are primarily fueled by his contention that military perception in warfare is comparable to civilian perception and, specifically, to the art of filmmaking. According to Virilio, therefore, cinematic substitution results in a 'war of images' or Infowar. Infowar is not traditional war, where the images produced are images of actual battles. Rather, it is a war where the disparity between the images of battles and the actual battles is 'derealized.' To be sure, for Virilio, wars are "no longer about confrontation" but about movement – the movement of "electro-magnetic waves" (Armitage, 2000).

John Armitage observes that

> similar to Baudrillard's (1994) infamous claim that the Gulf War did not take place, Virilio's assertion that war and cinema are virtually indistinguishable is open to dispute.

Yet Virilio's stance on the appearance of Infowar is consistent with his view that the only way to monitor cultural developments in the war machine is to adopt a critical theoretical position with regard to the various parallels that exist between war, cinema, and the logistics of perception (Armitage).

Like Virilio, both McLuhan and Debord declared war on the Spectacle of media. McLuhan is and was the ultimate media guru – so much so that in 1996 *Wired Magazine* mockingly canonized him as their patron saint. McLuhan's legacy as guest on *Laugh-In*, hipster, and "the oracle of the 'retribalized' Electronic Age" (Edmunds, 2008, 106) is perfect fare for today's YouTube generation. McLuhan's theories on media as an extension of our senses show that he was very concerned with the body as a technology, and his performative tendencies reveal that he also did not hesitate to translate himself into a character or to use his own body as a medium. Composer John Cage dubs him one of his oracles and uses him as source material for remixing in his aleatoric experiments with the *I Ching*, what Cage dubs "a 'discipline' that involved formulating a question then using coins to divine numbers that provide answers"; his oracles would provide the first letter for each line of the vertical structure (Funkhouser, 2007, 6). Critics have been less kind though for McLuhan's apparent lack of seriousness and McLuhan has been derided and ridiculed by them for this lowbrow behavior. In his article "Flip-Side Overlap: The Medium is the Music," Michael Edmunds demonstrates that Marshall McLuhan was a performance artist, and Edmunds takes this as evidence that McLuhan's "performance art has been poorly understood" (Edmunds, 2008, 106). Witness the interview he gave to Father Patrick Peyton on November 14, 1971 (McLuhan, 1971). The media philosopher says to Father Peyton that contemporary man:

> wants an interface, a resonant dialogue. He wants to rap, chat and empathize with everybody about everything, and this constitutes an interface of change in dialogue. It isn't just the passing of gossip back and forth. It is a kind of interrelating by which people feel that they are changed, that they are getting with it, they are getting involved, they are participating (McLuhan, 1971).

In our age of participatory culture, McLuhan's call to arms has resonance for us. While McLuhan and Father Peyton are discussing telephones and satellites, it is hard to believe that this discussion was taking place in a pre-personal computer world – and indeed McLuhan died the same year in which the first personal computer was released, so he never actually saw the PC revolution. But he saw it coming.

I have previously mentioned that performance art exists as a counter-cultural form on the basis of its unarchivability (another attribute of extreme perfor-mance) and its desire to insert bodies back into our consciousness as a medium of expression. Think, for example, of Jodi's viral invasions on your computer. Would you want to archive it, even if you could? Feminist theorists have tried to write the body back in as performative site. They do not see it as a passive object to be written on, but a process of trajective flows in motion that has become writeable, has become an agent of writing.[6] Performance art is an exploration of our sensual engagement with the world and with ourselves as objects-as-events in process in an isolated, acontextual moment in time.

Where subjects in motion in space create social networks with other subjects and with objects, the body as a cultural object creates its own connec-tions with other objects. We might think of this as equivalent to Bruno Latour's tagged world, the Internet of Things. Objects are our tangible interface with the sensory and virtual worlds, being windows on our own conceptual architectures of subjectivities. History too is a connective matrix born of the narratological impulse to link objects and events in space-time. Jay David Bolter and Richard Grusin argue that our contemporary culture is torn between the twin desires of immediacy and hypermediacy. We want our media to be transparent and invisible, but we also want more media (Bolter and Grusin, 2000, 5). As a result, the media are self-reflexive, multi-directionally mirroring the past and their predecessors even as they seek to reform, reuse, and restate them. This is what Bolter and Grusin call a process of "remediation" (2000, 19), an acknowledgment of Marshall McLuhan's observation that the media are containers for their predecessors, but also an extension of the medium itself as an embodied gesture in time. This signals that this interplay is not a one-way conduit: the performance of these attributes alters both old and new forms.[7] Network society reformats things according to the way in which we actually do things. And the nature of mobile media and handheld technologies makes us schedule time and space entirely differently.

The body writes itself in immersive space as a dynamic event or disturbance. This is not movement for an audience, but a dynamic and personal act that a performer performs for and by herself – which takes us back to Said's notion of the extreme, self-reflexive event. Performance is a form that has been favored by the disenfranchised, in part at least no doubt because difference, including race, gender, and sexuality, is so readily foregrounded and problematized in the body of the performer. As a dynamic and transitory art form, performance is ethereal, existing both inside and outside of time and space, inside and outside of representation and the real, inside and outside remembering. It is also readily misinterpreted. There are distinct links here with Chicano theorist Gloria Anzaldua's concept of a leaky border culture, where a new world emerges from the geographic convergence of two previously separate societies and races. In *Borderlands/La Frontera* Anzaldua (1987, 3–4) says:

> borderland is a vague and undetermined place created by the emotional residue of an unnatural boundary. It is in a constant state of transition. ... The only 'legitimate' inhabitants are those in power, the whites and those who align themselves with whites. Tension, ... ambivalence and unrest reside there and death is no stranger.

Anzaldua's writings address the borderlands/la frontera and cross-pollenating hybridities between Mexico and the United States, but we can see the shores of pirate activities also occupying such a marginal space. It is a site of disturbance and of seeing double, seeing again.

Constructing a 'second time,' the time of re-turn or re-visitation to the same material is not a repetition. It is a remix. Each re-presentation is a discourse network that operates within its own borders under a private system or logic, just as the real has its own "discursive and imagistic paradigms" that it creates and draws from (Phelan, 1993, 2). As we have already seen, remixes can take many different forms and they always interrupt time-based or sequential forms of telling. In a time before digital technologies became the be-all-and-end-all, spatialization was introduced into analog forms in other ways. For instance,

> In 1967 a revolutionary new way of seeing was introduced to televised sports: the instant replay. McLuhan was so enthralled by this effect that he dubbed ours "the age of the instant replay" (1973: 218). The replay was revolutionary because of the way it spatialized time-based events and, for the first time, allowed the viewer to derive the meaning of an event without having lived the experience (McLuhan, 1973: 219). A kind of flashback, it foregrounded the notion of seeing again or of recognizing the familiar in the new. McLuhan dubbed this "pattern recognition" (1965: 63) in sympathy with the ability exhibited by artificial intelligences to identify voices or visual repetitions (Guertin, 2008).

Peggy Phelan defines performance as "representation without reproduction" (1993, 3) because it always exists virtually, on a borderland, in the present moment on the verge of becoming. Phelan writes, "Performance's only life is in the present. Performance cannot be saved, recorded, documented, or otherwise participate in the circulation of representations of representations; once it does so, it becomes something other than performance" (Phelan, 1993, 146). The experience of the performance cannot be archived either. Just as our information is on the verge of a digital dark age, so it is the actual movement of the body in space at this moment that is the primary narrative here, as a break boundary that renders the mediation of the written text visible. For Jacques Derrida too, because of this present-tense element, performance is always already a kind of writing surface (quoted in Phelan, 1993, 149). As engaged audience members, we perform an act of translation with our eyes as we write ourselves into the surfaces and depths of the reading surface of the text as a performative event in the gap of the present moment.

Edward Said finds the way in which "extreme performance" resituates "an ordinary, mainly domestic and private passage of time, to an occasional, heightened public experience" (quoted in Bogdan, 2008, 2; Said, 1991, 10) to be of particular importance to the live versus mediated musical experience. In performance art, time too becomes a key element in how the body operates in public space. The role of the body in performance art, in fact, is as complex as the role of the body in memory. As with subjectivities, we all have several bodies (not a single one) that unite our complex sense of multiplicities and hybridities. These are social and cultural bodies – remembered, gendered, classified, racialized, and situated – that come together in these performative spaces to link the materiality of our real body with the image of this staged body performing the process of embodiment. Anzaldua calls the conflicting multiplicity of selves that the hybridized possess generates "interfacing." She explains,

> 'interfacing' means sewing a piece of material between two pieces of fabric to provide support and stability to collar, cuff, yoke. Between the masks we've internalized ... are our interfaces. The masks are already steeped with self-hatred and other internalized oppressions. However, it is ... the interface ... between the masks that provides the space from which we can thrust out and crack the masks (Anzaldua, 199, xv).[8]

The interface is a literal rendition of aestheticized haptic form as surface, and, in electronic spaces, form does not mean simply a surface tension. It is what we must touch and immerse ourselves in (like in our memories) in order to navigate the fluid spaces and times of the virtual world. For Anzaldua, these 'in-between' spaces, borderlands and interfacings, are therefore extremely fraught: sites of ambivalence, resistance, self-denial, and agency all at the same time (1990, xv). These in-between spaces are the space of video installation, like in Eija or Farocki or Hafez. Like in gallery space, in Debord's films, or in the work of jodi, McLuhan's life foregrounded art in a way that life itself cannot, since life is full of the gaps and ambiguities between ourselves and our interface with the world. Performance art is always aware of the tripart structure of seeing, being seen, and seeing ourselves through another's eyes. Performance art foregrounds these myriad internal/external bodies, equal parts fragile shell and hardened armor, as a part of living, enculturated practice in any space – real, virtual, and surveilled.

Dynamic data and augmented bodies

As I have already mentioned, McLuhan too was infamous for his performances. We might think of him as a one-person #occupy academia movement or as the poster boy for media performances. His manner was strange in the extreme. Douglas Coupland (2011) writes that McLuhan:

> tuned in and out of conversations with friends and strangers, and during classes he would ramble, seemingly unaware of those around him, clicking in and out of reality. Many people, when describing their encounters with him, say that with Marshall you had a few seconds to say your hellos or make your point, and after that he was back on Planet Marshall.

McLuhan himself seemed to feel that ever-present media buoyed him up and kept him company; he was plugged directly into human, electric, and spiritual networks. As a walking encyclopedia and the embodiment of live performance, he was both dynamic data and augmented body. He said,

> Most of the relationships between men are now invisible. The human bond, the electric instant bond around the planet, is invisible … which is not unlike the things we were taking about with relation to the mystical body, which is entirely around us and entirely invisible. It is hard to understand where the media starts and ends. The media is an all-encompassing service, and man seems to be an inadequate figure (McLuhan, 1971).

Inadequate figure or socially awkward son, McLuhan loved to perform. His riffs were musical in the extreme and bordered on the hypnotic. What he said was often rhythmic, hard to understand, and begged to be listened to over and over like a favorite song. Today we would call him a spoken word poet or a VJ of the medium of performance. Performance art as a form in opposition to the message of art spoken by rarefied objects is so far outside the established order that it exists beyond the realm of language. Through inversions of the public space of the text and the private space of the body, we might consider the fact that his performances were designed to create break boundaries – just like Debord's films, and Debord's strategies of *dérive*, *détournement*, and psychogeographic wanderings – in an attempt to reawaken our sensations. As the digital media have permeated our lives with greater and greater ubiquity as McLuhan prophetically foresaw, so too this notion of immediacy – or "liveness", where liveness is a mediated event brought into existence by recording technologies – has moved to the front and center of contemporary arts.

Augmentation may well be one answer to how we reconcile our desire for the live with a technological cultural framework that delivers liveness. As the World Wide Web and image-based media make us more and more aware of the aesthetic attributes of information space that surround us, our conceptualization of ideas, histories, and stories becomes increasingly visual. We have always yearned for tools with intelligence, and many different kinds of software, engines, and technologies now undertake conceptual modeling that so far exceeds our own capacities that they render multilayered complexity visible instantaneously. This is a by-product of the massive changes that the world underwent in the Twentieth century. Three aspects of that transformation are particularly relevant to remix culture and digital storytelling. They are, first, an overabundance of events (information richness or overload); second, spatial complexity or the layerings of multiple narratives and histories over single places; and, third, the personalization of data, a.k.a. the folksonomy and the arrival of DIY culture and participatory cultures with their mobile technologies (Augé, 2008, 33). This combination of rich data flow, spatial multiplicity, and personal media has transformed everything from how we eat to how we love to how we communicate to how we shop to how we learn. Story is, of course, the space where we make meaning, the place where we connect the dots to form a thread of understanding that is our dynamic engagement with the world. The acceleration of history has met the network of connectivity that is globalization head-on, and that multilayered pile-up is the acculturation of storytelling, or mode of navigation through information space. In the Twenty-first century, this acculturation – the assimilation of a different culture – forms the building blocks of new kinds of narrative architectures. In short, acculturation is to place what augmentation is to space.

As information moves off our screens and into the world, it is apparent that the next big thing will be augmented interfaces. When Douglas C. Englebart conceived of the personal computer as a tool for the augmentation of the human, he foresaw an expansion of the intellect by virtual means, not as an architectural addition to real space. Virtual space and augmented space are distinctly different and it will be helpful to define the difference between the two. The virtual creates an immersive, simulated world (or place) that we interact with like the mainframe in *The Matrix* or the holodeck in *Star Trek*; augmented reality is, by contrast, a layering of dynamic digital data over real space like the Terminator's readouts on his viewscreen. VR and AR are not necessarily mutually exclusive, as Lev Manovich points out, since the difference between immersion and augmentation can be a question of scale determined by the size of the display (Manovich, 2006, 225). The real time combination and interaction of real and augmented data is not a new concept – head-mounted augmentation has been the standard interface in military mechanics since the 1990s, for instance. Mobile technologies that have emerged since 2008, however, are now putting locative media, augmented data mapping, and social media tools within reach of the networked individual, the media tactician, and the artivist.

Dynamic data surround us every day in just about every aspect of our lives. Video surveillance, locative or wireless media, and computer and video displays are entirely ubiquitous. At the same time as urban landscapes are information rich, they are networked, layered with multiple histories that can be mapped along racial, gendered, geopolitical, and cultural margins. There are two different ways in which these locative information flows can be mapped: (1) Annotatively, leaving scribblings on the surface of specific information spaces; or (2) Phenomenologically, tracing the movement of subjects in the world (Tuters and Varnelis, 2006). These are the information trails that comprise psychogeographic space. With SmartPhone apps urban wandering might be used to create personal, layered, or augmented psychogeographic narratives.

According to Guy Debord, a psychogeography is "the point at which psychology and geography collide, a means of exploring the behaviorial impact of the urban place" (Coverley, 2006, 10). Debord saw in psychogeographies the potential for counteracting the numbness that the spectacle of mass media impose on us – and locative media has been widely criticized for its very lowbrow accessibility paradoxically coupled with ties to the military-industrial surveillance network of GPS satellites combined with its tethering to consumer electronics. Locative media have tremendous potential for political activism precisely because of their very ties to the specificity of place, the ownership of the airways not withstanding. In fact, locative media open up the possibility of rewriting the imperialist maps of dominant cultural codes through guerrilla tactics. We can think of psychogeographic engagement as participatory culture, a smart mob of sorts that eliminates the audience (*à la* Alan Kaprow) and reinserts us back into formative story spaces as activists, authors, and interactors of our own stories. The importance of psychogeographies lies in their transformative nature. Annotative projects perform the Situationist act of *détournement,* "changing the world by adding data to it," and phenomenological ones perform the *dérive*, embedding socially aware knowledge – in this case stories – in data (Tuters and Varnelis, 2006). By definition, psychogeographies render walking an act of subversion that engages with urban space with the goal of reinventing or rewriting the city. They incorporate navigation with simultaneity – layered times past, present, and future – at geospatial coordinates. They are a political act of playful trickery in nonlinear space that is transformative to the creator/user. Next gen remixes and interactive stories will be layered, interactive architectures in real space that are visible, audible, and readable on SmartPhones. They offer potential for media users to lose themselves in the city while reconnecting with the urban environment as a way of taking the activist stance of owning their own stories and local histories. Locative media are inseparable from code in many ways, Tuters and Varnelis (2006) claim, because it requires programmic acts, including:

> turning individuals into processors. Virtually all locative media projects rely on programs for their execution. The resulting product is generally either delivered live to a user in the field who then performs the piece or, alternately, crystallized as an indexical trace of the event, later displayed at a gallery or on a website.

Since any code is language-driven, it can potentially become an empowering, generative tool in a way that a piece of software is not. Beyond the specifics of this application, clearly augmented fiction as a political art and form of critique has potential widespread uses far beyond our imaginings.

If remixing is how we make meaning, narrative is where we live. It is a landscape experienced through, as Simon Schama would say, "the archive of the feet." This landscape is also full space. It is shot through with embodied trajectories and:

> the potential but invisible field of possibilities nourished by everyday perceptions, lived experiences, different histories, narratives and fantasies. In fact, any understanding of [the] landscape [of story] entails a succession of distinct moments and different points of view. The layeredness of landscape, in other words, forms part of our own projection. Every landscape is also a mental landscape (Puranen, 2009, 11).

Mobile technologies raise the possibility, when combined with geospatial data, of making these mental landscapes visible. Their potential as a vehicle for remixing and storytelling is profound.

The combination of locative media and resituated storytelling has a venerable tradition. Practitioners of *Ars Memoria*, the ancient art of memory, in classical temples and pilgrims embarking on holy quests via pavement labyrinths in medieval cathedrals were enacting precisely these kinds of narrative journeys. Our technological architectures have changed somewhat in the intervening years and locative media for us carries different implications. The work of Vancouver artist Janet Cardiff (in conjunction with her husband George Bures Miller) is one such example. Similar to early text-based computer games that prompted the user to search the environment with commands such as "look right," "turn left," and "open the mailbox," Cardiff's "audio walks" direct a user through a real urban environment. Via an audio player, headphones, and movie clips, the walk is interlaced with real experience, historical references, snippets of narrative, virtual "sound effects and other aural data" (Manovich, 2006, 226). For instance, in "Ghost Machine", the interactor winds their way through a maze of backstage staircases, normally off limits to the public, trying to find a man hiding in the attic of Berlin's beautiful, historic Hebbel Theatre built by Oskar Kaufman in 1907. The narrative is unclear, but eventually it becomes apparent that the interactor, discovering herself on stage, has been a participant in a play all along. This is history as remixing or archival data capture as story. The technologies and ideas are simple, but the experience is profound owing to the immersive nature of the visual and audio experiences in combination with the meeting of the past of the recording and present of the experience (Manovich, 2006, 226). The listener thereby becomes an active participant in a psychogeographic narrative space. Similar experiments are being done with virtual museum tours and urban walks, but the emotional impact of Cardiff's art is in the narrative of place or the landscape of the feet. For anyone wanting to evoke that ghostly presence which continues to haunt the city and its residents, one of the most chilling and effective SmartPhone augmented reality apps is one that re-erects the Berlin wall on your phone's screen, allowing you to retrace its path.

The original wanderer in city spaces, the *flâneur*, was an urban soul who emerged out of the new cityscape of the early Twentieth century. He was a ghostly wanderer who transformed his surroundings, and who was at one and the same time an outsider and a part of the crowd. For the *flâneur*, navigating the city was a visual act and a form of creative expression. Getting lost was more artful still. It was urban development and the rise of consumer culture that eventually drove the *flâneur* out of the city. Replaced by the more productive shopper, the consumer laid little claim on the space of the city itself – her eyes were always glancing toward the next product or commodity. The consumer's death-knell in turn was really rung by the arrival of a different kind of machinery: not spinning economic wheels, but the automobile's. Where changes in the cityscape, modes of travel, and data acceleration seemed to threaten to make the pedestrian redundant, augmented reality technologies could establish a new kind of wandering in the many layers of real time in urban space for the exploration of narrative space. Right now, there are many augmented reality apps available for SmartPhones that use GPS technology to lead you to the nearest glass of Stella Artois or help you find that million-dollar condo in Paris that you've been yearning for. Disney already made liberal use of the technology in a poster campaign for *The Prince of Persia* and in an interactive marker-based application for the film *Clash of the Titans* that allowed users to "Release the Kraken" on their own computer.

How might this richness of place be united with urban wandering to create or retell new or familiar or layered or augmented psychogeographic narratives? Debord saw in psychogeographies the potential for counteracting the numbness that the spectacle of mass media impose on our senses. In contemporary terms, psychogeographic engagement is not dissimilar from participatory culture – a culture that eliminates the audience. Kaprow in his 1965 work entitled "Notes on the Elimination of Audience" explores his invention of 'happenings,' artistic events in which the audience participates. These events were intended to create a heightened experience of the everyday where interactors could fuse with the space-time of performance. He advocated that all audiences should be eliminated entirely and individuals should become participants. Not to be confused with theater or performance art, Kaprow's happenings were improvised in the moment like children's imaginative play while following the parameters of a predefined script. Digital technologies could enable this kind of engagement with a location or event in personal, virtual, psychogeographic ways.

Some technologies are already in existence that point the way toward embodied, locative stories. Within our increasingly DIY culture, existing iPhone/iPod/iPad apps (including *PlaceThings, Layars*, and *Wikitude*) might be adapted, repurposed, or developed. *SixthSense* is a prototype developed by Pranav Mistry and Patti Maes at the MIT MediaLab. Based on a theoretical technology they designed for Steven Spielberg's *Minority Report* in 2001, *SixthSense* is assembled from off-the-shelf technologies, including a camera, a data projector, and a cell phone. Together they generate real-world interactive interface that will search, capture, store, analyze, and project. This is a tool for generating data rather than narratives, of course, but new kinds of narratives could be extrapolated from its findings. *PlaceThings* is a

SmartPhone app developed by Dean Terry at the University of Texas at Dallas. *PlaceThings* allows users to deposit and read geotagged data in real space. A kind of virtual graffiti, *PlaceThings* becomes a vehicle for them to comment on the technologies they are helping to create. *Wikitude* is another iPhone app that combines Wikipedia data with locative media to generate local 'histories' or perhaps more accurately a tool for navigating local data in real space. *Layars* is the most advanced augmented reality SmartPhone app and the most promising. Using sophisticated GPS tagging, it enables users to intcract with virtual objects in real space. Working with the new mobile apps, before too long we should be using data capture, data bodies and augmented selves to conjure up characters and perform live stories in air.

Authorship

From karaoke culture to vernacular video

In a recent talk for TED, bad-boy composer, artist, and performer Malcolm McLaren lambasted contemporary art practitioners for their lack of authenticity. He says that they have abandoned the messiness of process in what he calls today's Karaoke Culture. Karaoke is, of course, the form popularized in Japan that has amateurs performing pop songs via teleprompter in pubs all over the world. It is not so much the amateur nature of Karaoke Culture that McLaren seems to sneer at so much as at the lack of 'originality' and the lack of end-product. Musical impresario, punk provocateur, perhaps as famous as the manager of the Sex Pistols as he was for being a performer in his own right, McLaren reveres the "messy process" of creation. No doubt he would find few or no redeeming qualities in someone like Girl Talk, the club name of the sampler and mashup musician Gregg Gillis. Girl Talk came to fame remixing bits and pieces of existing works to create sampled 'new' works that people dance to, live, in clubs. More surprising though is McLaren's sneer, since McLaren himself was known for his hip-hop, electronica, and opera-inspired works. Part of Girl Talk's compositions' charms lie not simply in the familiarity of his resituated and reimagined material (as in the case of McLaren's visionary work), but in the copying – the exact repetition – of the known. Of course, Girl Talk's material cannot be sold due to copyright restrictions, so instead his income stream comes largely from his performances.[1] But if Girl Talk's art does not consist of the messy process of creation, then what is it? His preconfigured loops and refrains are prepared in advance, but assembled live. Gillis as VJ/programmer is neither originator nor author. The end-product may be made up of pre-existing fractions of seconds or loops of other people's work, but the autonomous, live, performed experience is all Girl Talk.

California new media theorist and artist Eduardo Navas defines the cut-and-paste practices of remixing as having three different forms. The first form is extended remix. It was designed to create longer, more danceable versions of instrumental sections of songs. The second form is selective remix; it adds and/or subtracts material. Again, this is a practice most often associated with music. The third form is what concerns us here. Reflexive remix creates a unique and original work from pre-existing parts. The Frankenstein monster of the remix, it acquires a life of its own, acknowledging its (recognizable) source(s) and reflecting back on its own act of creation. According to Navas, it is this conversation with its past incarnations that makes it a true remix: "it allegorizes and extends the aesthetic of sampling, where the remixed version challenges the aura of the original and claims autonomy even when it carries the name of the original; material is added or deleted, but the original tracks are largely left intact to be recognizable" (Navas, 2011). This process of resituating the familiar adds a metatextual layer to the work, and the use of transformed but readily identifiable components is what distinguishes this form from something that is merely derivative or plagiarized. The conversation with itself and its origins arises out of this apparent history. Appropriation art has used similar methods, of which Sherrie Levine's remixes of Edward Weston and Walker Evans are perhaps the most famous.

Celebrated and controversial conceptual photographer, painter, illustrator, printmaker, and sculptor Sherrie Levine has for many years used her art to challenge the patriarchal establishment of the art world and the conventions of spectacular images. Fascinated by the premises of originality and authorship as a critical framework for her practice as a – by definition – alienated woman artist, she has always tested the boundaries of traditional representation in her art. She has done a series of appropriation works since the late 1970s that come 'after' the original artist. The titles of these works all follow the formula of "(Untitled) (After [insert Name of Artist])," and include "Edward Weston," "Walker Evans," "Alexander Rodchenko," "Franz Marc," "Ferdinand Léger," "Piet Mondrian," "Joan Miro," "Paul Matisse," and "Paul Klee."[2] Some of Levine's works are photographs of the original photographs, and others are collaborative photographic collages of the original authors' works. Levine does not see her work as derivative, but as regenerative instead. For her, her intervention in the original is an act of collaboration that transforms the muse work into a symphony or a conversation with her present-tense intervention. For instance, her "After Edward Weston" photograph is a photograph of Weston's photograph of his son posing as the ancient Greek statue of Hermes carved by the greatest of the 4th-century B.C.E. Attic sculptors, Praxiteles. In actuality though, none of Praxiteles' own statues is known to have survived, but many (often Roman) copies did. So Weston's re-creation of the statue using his son is a copy of a copy in the first place. Douglas Crimp observed about Levine's photograph,

> According to copyright law, the images belong to Weston, or now to the Weston estate. I think, to be fair, however, we might just as well give them to Praxiteles, for if it is the image that can be owned, then surely these belong to classical sculpture, which would put them in the public domain. … The priori Weston had in mind was not really in his mind at all; it was in the world, and Weston only copied it (Crimp, 1980).

So Levine's exploration of the image becomes a recursive figure spiraling back through spheres of influence across centuries of art history. In fact, there are so many so-called forgeries that exist in art because copying in the days before museums and galleries used to be the only way to gain access to art – all of which was patron- or Church-owned.

Michel Foucault, in his famous essay "What is an Author?", says that the author is never the origin or location of meaning. Meaning instead is fluid and changeable, always situational over time. Each work exists in conversation or in quarrel with other works. All works, he says, comprise quotations from other works, and "all discourses are objects of appropriation" (Foucault, 1986, 108). The author/artist herself is a huge repository of pre-ordained knowledge from which she draws, and every book is a rich network of influences, connections, and imitations. These forces undermine any notion of authorial intent and so in her absence, he argues, we should understand the original work to be as problematic as any notion of author. In response to Weston as patriarch, we can see that Levine deliberately undermines his authority both by coming 'after' and by signing 'her' name to the new work. Craig Owens, in "The Discourse of Others: Feminists and Postmodernism," observes:

> [W]hen Sherrie Levine appropriates – literally takes – Walker Evans's photographs of the rural poor or, perhaps more pertinently, Edward Weston's photographs of his son Neil posed as a classical Greek torso, is she simply dramatizing the diminished possibilities for creativity in an image-saturated culture, as is often repeated? Or is her refusal of authorship not in fact a refusal of the role of creator as "father" of his work, of the paternal rights assigned to the author by law? (This reading of Levine's strategies is supported by the fact that the images she appropriates are invariably images of the Other: women, nature, children, the poor, the insane….)

Levine's disrespect for paternal authority suggests that her activity is less one of appropriation: she expropriates the appropriators (Owens).

By shifting the authorial power however, Levine both shares authorship and refuses it. We can see Levine's act of re/possession in her bronze "Fountain/ After Marcel Duchamp." Duchamp's original readymade has been lost and all of the extant copies of his 1917 work that are in galleries and museums are replicas which he produced in the 1960s from a photo of the original by Alfred Steiglitz. Levine's copy of a copy of an absent original, therefore, functions very much like her version of Weston's Hermes after a lost Praxiteles. Duchamp likely would have been amused by the clash of highbrow and low cultures that meet in her bronzed urinal. Ray Dowd says that Duchamp derailed copyright law because it was an artwork with a utilitarian function. As a commercial object it had no component of expression that was protectable under the law. Duchamp himself believed that readymades should only have a half-life of 25 years (as opposed to 70) in the copyright sphere – "and that it could not – indeed should not – outlive the artist because the original context in which the work was conceived would have changed so much that the work would effectively be rendered meaningless" (Sahiner, 2011). Levine made her point, but lost that battle, at least in the case of Walker Evans' photographs. Even though his works were out of copyright and allegedly in the public domain, the estate sued and won rights to Levine's photographs. Then, in 1994, the estate of Walker Evans was donated to the Metropolitan Museum of Modern Art in New York; his estate included Levine's photos and MoMA still hold the rights (Jana).

Levine's genealogy of appropriation does not stop with her photographs. Her work in turn has been appropriated by artist Michael Mandiberg, who digitized her copies of copies – works that were made available by MOMA for this project. He could have paid $100 to $300 per image to make his project legal, but on behalf of Levine – or on behalf of art in general? – he chose not to (Jana). He now offers his copies of her copies as printable objects with certificates of authenticity.[3] His works exist at two recursive and identical websites: one is aftersherrielevine.com and the other is afterwalkerevans. com. In essence, Mandiberg's readymades (as he refers to them at his website) are open source versions of the images that anyone can duplicate. In addition, in a particularly Escherian turn, Levine too was not content to leave her recursive series there either. For the 2007 Venice Biennale, she revisited the series and exhibited a new set of digital 'afters'. Her "After Paul Cézanne" is a sequence of 18 blow-ups of pixels shown as prints taken from a digital version of the original – a translation of a translation? Her "After Steiglitz" is a series of 18 digital, pixilated images that copy his "Equivalents" series from the 1920s and 1930s. Steiglitz's was a very significant series because they were deemed to be the first nonrepresentational photographic images. The original images, derived from clouds, were the first photographic abstractions. Levine's copies of Steiglitz's abstractions, therefore, become subjective digital captures of recursive representations of the original nonrepresentational image. What becomes of representation when even abstraction itself is a subject and object to be copied? According to French theorist Nicolas Bourriaud art is no longer the termination point. Instead the artistic process is about *postproduction*. The goal is to start a conversation and it is in the process of conversing that art is born. No longer is there an original object to bow down to. Instead, all participants are potentially equals in the conversation. Bourriaud (2002a, 19) says,

> the contemporary work of art does not position itself as the termination point of the "creative process" (a "finished product" to be contemplated) but as a site of navigation, a portal, a generator of activities. We tinker with production, we surf on a network of signs, we insert our forms on existing lines.

Appropriation art, as Levine practiced it in the 1970s and 1980s, has in turn been gradually eroded and replaced by a new school, the objectless relational aesthetics, which are social, interactive collaborations. The kind of collaboration that relational aesthetics fosters is exemplified by artist Rirkrit Tiravanija's two-person show *with* Andy Warhol. Born in Buenos Aires of Thai descent and currently residing in New York, Tiravanija has won the 2010 Absolut Award and the 2005 Hugo Boss Prize offered by the Guggenheim Museum, along with many other honors. The press release described his Warhol retrospective as follows:

> In 1994, Tiravanija made/curated a two person show with his other half, Andy Warhol. It was a hybrid retrospective of sorts for each artist. Tiravanija created a binary set up of three pairs of work, with one work by each artist in each pair: A Mao and a stack of beer bottles; a Brillo box and a wok; a bed and a pile of books and movies. Each pair created a metaphysical and cultural bridge across time and space from one world to another. Each side looking at the other in the mirror and being disgusted at themselves. One side surface and mediated, the other dirty and touched, but both steeped in melancholia and necrophilia. (Vernissage TV, 2011)

The practice of social sculpture has even found some footing in China as with Ai Weiwei's event at Documenta in 2007, when he invited 1001

Chinese nationals to Kassel as a "socio-political readymade" (Münter, 2007). Where Levine's audacious act of appropriation had blown notions of originality out of the water and forced us to re-examine authorship, mediated pop art – complicit with the numbing spectacle – set consumer culture up as an abstraction, as a theme even, for us to contemplate. With Warhol's pop art, there was no outside and the act of representing a stolen image (although, in Warhol's case, he always paid for the rights to use the image) was supposed to be an ironizing comment on consumer culture. In actuality though, representation immediately commodified the pop art critique and any ironic distance was lost. Bourriaud argues that in the early 1980s artists like Barbara Kruger critiqued the simulation of pop art, which set up the representation as an "absolute commodity" where "creation" was "a substitute for the act of consuming" (Bourriaud, 2002a, 26). Kruger's "I buy, therefore I am" sought to address this problematic lack of an outside in the framework. Tiravanija's relational aesthetics-based recontextualization of Warhol's work is a non-consumer culture framed exposé of Western assumptions. In the shift, it is clear how there is no longer a one-to-one correspondence between works for Tiravanija as there had been for Levine. It is no longer just enough to start a conversation like the (hidden) message of gender relations and copies without originals in Levine's work. Tiravanija introduces the genuine objects of Asian life that are pregnant with history, and use-value, which underlines their lived-in-ness, their used nature, and foregrounds how steeped these items are in Eastern scarcity and non-spectacularized politics. Warhol's Mao and Brillo Box by contrast become commodified symbols of highbrow art. Tiravanija's simple objects are stand-ins for necessities. Where we are now, in a postproduction period, is "beyond what we call 'the art of appropriation,' which naturally infers an ideology of ownership, and moving towards a culture of the use of forms, a culture of the constant activity of signs based on a collective ideal: sharing" (Bourriaud, 2002a, 9). Significantly, Tiravanija's work too has moved on into cooking-based installations and sensory immersions, where the shared meal becomes the aesthetic experience.

It is the experience of the act of sharing creation – the process of circulating – that has become the aesthetic of production for the multitude in a digital age, and this is what the copyright barons and prosecuting lawyers do not understand or do not choose to acknowledge about the legitimacy of the use of appropriated materials as building blocks in creative works. French film director Jean-Luc Godard had greater insight than McLaren when viewing the new culture to be found on the Web. Recognizing that the form is not a visual one, he said, "people make films on the Internet to show that they exist, not in order to look at things" (quoted in Lovink, 2008a, 9). This is a big departure in the shift from a visual to a dynamic or social culture. A YouTube clip, for instance, is a piece of information posted on the Web that is designed to kickstart a conversation. This is what Pierre Bourdieu would call cultural capital. It is in the social engagement with a larger cultural framework that the form springs to life, acquires value, and fulfills its purpose as a social agent of personal identity. These so-called objects are more accurately acts of self that perform the tri-part formula of interruption, disturbance, and capture/leakage. First and foremost, they catch our attention – with the interruption – often by using familiar formula, genres, or material. Second, they enact disturbance through the buzz of conversation and viral responses they initiate, and third, they capture the familiar and retransmit or leak it through a social media interface.

YouTube is, in 2011, ranked globally as the number three most visited website after Google and Facebook. It was a primary mover in fostering the first big wave of lowbrow video cultural formulation and so it makes sense to take a closer look – even if most of its contents are drek through and through.[4] Established on December 15, 2005, YouTube's primary goal, as social media, is the reciprocal sharing of information. YouTube uploads might comprise reposting clips from other cultural sources, sharing jokes, or demonstrating personal abilities or achievements. In an exploration of YouTube as a mimetic form, Birgit Richard outlines the community's seven basic genres of undertakings. These are response pieces, cover clips, remakes/adaptations/interpretations, parodies, remixes (which include found footage, television, movies, premakes,[5] 60-second or five-minute versions of full-length feature films or performances, mashups, games, cartoons, music videos), re-enactments, and ego clips. Ego clips are by far the largest category and include everything from pop star Justin Bieber's karaoke videos (which gave him his start) to sporting accidents to personal travelogues to self-promotion of all kinds – including viral marketing pieces like "Where the Hell is Matt?" Other distinct kinds of ego clips are fan material, anti-fan or hatred postings, flamewars, witnessing (like the death of Neda Agha-Soltan during the election riots in Iran in 2009, considered to be the most widely viewed death in human history, or mockumentaries as a form of commentary on other genres and topics), experiments, and demos (electronics jailbreaking, dynamic typography, etc.). The greatest innovation and artistry are generally seen in the categories of remixing, fan materials, and art response.

Part of these videos' cachet is in the fact that they operate outside of the economy of exchange and often do so with appropriated – or shared – materials. At first that was how YouTube worked and it was relatively untroubled by the copyright enforcers. That changed when Google bought the 11-month-old YouTube in 2006. The DIY and appropriation aesthetics were eroded as Google went head to head with Viacom in a $1 billion lawsuit.

Google won, but won by selling its soul. It agreed to the implementation of a copyright filter, which has been applied since October 15, 2007. Now, account holders are regularly contacted by corporations regarding alleged copyright violations, and videos can be removed or soundtracks (and accounts) deleted without the poster's permission. YouTube uses a manual copyright takedown notice under the Digital Millennium Copyright Act (DMCA) system, delivered via a copyright complaint form – available to anyone who wishes to request the removal of a particular video. This generally happens without discussion, only with a statement of a violation, and is done without attention to the rights to enforce the law, or exceptions to the rules like parody or fair use. For example, of the top ten viral videos TIME selected for 2008, three have since been removed for alleged copyright infringement. YouTube also introduced an Audioswap service that allows for the automated replacement of a restricted soundtrack with an acceptable one from YouTube's media partners. At the same time, YouTube made Content ID, a state-of-the-art search engine, available. It is a tool for copyright holders that is supposed to allow them to 'control' their content better. It packs two punches: (1) It enables the owner to "identify user-uploaded videos comprised entirely OR partially of their content" in both audio and video tracks, and (2) It allows the owner to "choose, in advance, what they want to happen when those videos are found" (http://www.youtube.com/t/contentid). Corporations or artists can remove videos or leave them up and cash in, using them as marketing opportunities to profit from 'infringement' or fair use, to track and get stats on them or block them altogether. Designed to generate revenue for the corporate media producers, this program is supposed to alleviate infringement issues. Effectively, though, what really happened is that essentially YouTube changed sides. Instead of providing a venue for individuals to upload their stuff and 'broadcast themselves,' they now have a site that serves the media conglomerates' hunger to cash in on remixed content. In addition, videos are now *tracked.* YouTube searches them for parallel or repetitive material, responses (that is to say hits) and viewing patterns, along with regional, demographic, and locational (where is the video embedded?) stats through a utility called YouTube Insight. Insight compares the masses of files that are uploaded every minute against known, existing content, and creates a spectogram or fingerprint to compare against other files. The spectrogram uses fuzzy logic, which means it can identify similar objects that are coded differently. When matches are found, Insight follows the algorithm stored for that site: delete audio, delete video, keep audio, or prevent one piece of audio from being associated with a particular video or leave it unchanged (Van Buskirk, 2008). Second, the database rules are always in a constant state of revision as laws change, rights change, and labels change hands. This means that

Insight checks not just new uploads, but all existing content at the site. One of YouTube's software engineers, Mary Gould Stewart, claims that

> YouTube is empowering content owners by providing a culture of opportunity, where "progressive rights, management and new technology offer an alternative to simply blocking all re-use of intellectual property. YouTube's approach, she argues, sustains new audiences, art forms, and revenue streams that would otherwise be circumscribed" (quoted in Clay, 2011, 221).

She fails to mention that services like RoyaltyShare are ignored, and the uploader of remixed or derivative material can *never* collect royalties. In addition, Stewart does not point out that all of this is in strict violation of fair use laws, which corporations are encouraged (through YouTube's engine) to ignore. Fair use laws are pretty much under siege everywhere and for the most part keep losing ground in the courts.

The most visible and widely protested example of the abuse of these rights by corporations are the parodies of Oliver Hirschbiegel's critically acclaimed 2004 film, *Downfall* (*Der Untergang*). One of the most often imitated pieces of content on the Web in recent years, one particular scene in the film has had more than a thousand responses to it uploaded to YouTube. I will discuss a few of these at length later in this chapter, but the controversy in this instance starts because the parodies do not just reimagine the scene but actually make fun of it. The scene that has been so lovingly rewritten so many times is Bruno Ganz's rendition of Adolf Hitler as he blows a gasket when his generals inform him that the war is lost. The popularity of this as remix material no doubt stems in part from the fact that the only changes the parodies make is that they alter the subtitles – and alter them so that they follow what Ganz and his generals say as closely as possible. As a low-tech entry point into the art of remixing, this clip offers something for everyone. These parodies show Hitler blowing up over many different slights, insults, and omissions. Hitler loses it when Michael Jackson dies, when the iPad is revealed not to have multitasking, when Hitler learns of the *Downfall* parodies, and when Hitler learns that Constantin Film, the copyright holder, is taking down the *Downfall* parodies. In 2010, Constantin asked YouTube to remove all of the parodies. That the parodies were unauthorized is without question. Does Constantin have the right to remove them, even if they are harming the film's reputation or taking too much of Constantin's energy to monitor? From the beginning, Constantin has issued 'cease-and-desist orders' under the DMCA for an "embarrassing appropriation," according to Andrew Clay, and they have also been hit with complaints about the other products or technologies that have been satirized – especially for the Microsoft Xbox Live version, which received more than four million hits, and protested Microsoft's habit of shutting down Xboxes for infringement of copyright (Clay, 2011, 225). Constantin has been heavily criticized for its move to suppress the parodies. YouTube's unprecedented creative energy is in jeopardy from this abuse of the system. The Electronic Frontier Foundation (EFF), an American nonprofit for civil liberties of digital technologies and online use, claims:

> The life blood of much of this new creativity is fair use, the copyright doctrine that permits unauthorized uses of copyrighted material for transformative purposes. Creators naturally quote from and build upon the media that makes up our culture, yielding new works that comment on, parody, satirize, criticize and pay tribute to the expressive works that have come before. These forms of free expression are among those protected by the [American] fair use doctrine (Clay, 226).

In fact, even the Chairman of the EFF made his own *Downfall* version, which I will discuss later in this chapter. Many of the videos that have been challenged have now been returned by following the correct channels to appeal YouTube's removal of the content, and many more new videos continue to be uploaded. With the hand-over from Viacom, Google also became the owner of all clips – which, at the very least, is a major discouragement to artists wanting to post their work. In 2008, to counter this new policing trend, to increase revenue, and to compete with other new startups like Hulu, Google reached a deal with MGM, Lions Gate Entertainment, and CBS to broadcast their full-length shows through the site. This makes their pieces viewable, but it does nothing to foster or compensate the creative output of YouTube users.

Value (or cultural capital) in the social media universe is garnered through stickiness. As a mimetic form, the more a piece is discussed, commented on, imitated, parodied, and remixed, the more successful it is. Initially the site featured mostly videos that were captures of pre-existing material. Over the past few years, YouTube's greatest successes have been in the crowdsourcing of ideas, and at creating "new hybrid forms of production and reception" (Richard, 2008, 149). In its intermediate years it has become a cradle that supports the creation of emergent art forms. As Bill Seaman has noted, the Web has a 'recombinant poetics' and, as Tom Sherman says, part of these principles of recombinatory works is the constant revisitation of pop culture as an act of "repetitive deconstruction" (2004, 161). It is this cycle of constant revision that makes video stand apart; it is neither an art form nor "a critical method" (Mitchem, 2008, 278). It is instead "a critical practice"

(Mitchem, 2008, 278). Think, for instance, of Perry Bard's *Man with a Movie Camera: The Global Version* and how uploaded clips converse with other clips as well as with Vertov's original. Bringing together televisual and cinematic and other references with clipart,[6] comics, and game frames, for example, vernacular video celebrates its low-resolution attributes (known as YouTube quality) as a part of its aesthetic.

What is likely the largest crowdsourcing project to date was sponsored by YouTube and *National Geographic.* It is director Kevin Macdonald's *Life in a Day*, a documentary produced by Ridley Scott that premiered at the Sundance Film Festival in 2011. Oscar-award-winning Macdonald, best known for *Last King of Scotland*, collected 80,000 entries from around the world all shot on the same day – June 24, 2010 – and, along with editor Joe Walker (of Steve McQueen's *Hunger* fame), distilled them down into a feature film that runs for just under 95 minutes. Setting himself the impossible task of making a movie all around the world on the same day, Macdonald's inspiration was:

> a British group from the 1930s call the Mass Observation movement. They asked hundreds of people all over Britain to write diaries recording the details of their lives on one day a month and answer a few simple questions…. These diaries were then organized into books and articles with the intention of giving voice to people who weren't part of the "elite" and to show the intricacy and strangeness of the seemingly mundane. I simply stole the idea! (quoted in Watercutter, 2011).

While Macdonald's intent had initially been to celebrate the fifth birthday of YouTube, he wanted to "take the humble YouTube video, long considered the dregs of cinema, and elevate it into art" (Dodes, 2011). Macdonald has a whole new appreciation for YouTube now. Before, he saw only a site where people recycled materials. Now, he sees a participatory interface and a wellspring of creative energy: "The other side to it that we've tapped into … is this extraordinary creative community of people who are spending a lot of time making films" (quoted in Dodes, 2011).

Although Macdonald does not mention them, the work may also have been inspired in part by the work of Florian Thalhofer and Katarina Cizek. Thalhofer's nonlinear documentary *Planet Galata* (created with Berke Bas) is about Istanbul's Galata Bridge and the homes, people, businesses, and worlds that inhabit it. Thalhofer defines linear film as having been the result of "technical limitations during pre-computer times" (http://planetgalata. com/). Cizek's groundbreaking documentary projects at the National Film Board of Canada have been particularly innovative in terms of thinking about new multi-perspectival approaches to filmmaking. They include her interactive webwork *Highrise*. *Highrise* was "a multi-year project to document the human experience in vertical suburbs" and parallel to it she ran an experiment with a participatory component in Toronto using six tower-dwellers as subjects shooting their own perspectives on the world from their windows. That became a collection called *The Thousandth Tower*. Then she took the project global as an interactive piece that resulted in Cizek's Emmy Award-winning film, *Out My Window* (2010; http://interactive.nfb.ca/#/outmywindow). The *Out My Window* website has a 'participate' category where people can upload their own narratives about and photographs of their elevated views on the world from their own apartment towers. Meanwhile, working with architects, the original project has been expanded in *The 2000th Tower* where people imagine improvements to their neighborhoods and animators render the ideas. *Out My Window* also has an interactive media lab installation at the NFB headquarters where users can explore the work in real time and space.

Using YouTube's upload function, Macdonald's *Life in a Day* team amassed 4500 hours of footage from 192 countries. All of the chosen contributors are listed as co-directors. Multilingual film students were hired to sift through the footage, and 25 people were employed full-time for two and a half months to view the material and sort it into categories, including countries of origin and themes, and to assess it on a six-star system for quality. The technical hurdles themselves were complex, as, for example, they received material shot at a staggering 60 different frame rates that all needed to be converted. Most of the footage had been freely submitted, but 400 cameras were also sent out by mail into the developing world to ensure that the whole planet would be represented; those submissions were returned on memory cards also by mail. Significantly, 25 percent of the final footage used was from this mail-sourced footage, from people who had no access to YouTube. Crowdsourcing has allowed for the creation of an entirely different kind of film, one that would not have been possible without the technology. Like Thalhofer's and Cizek's work, *Life in a Day* has received critical acclaim for its real time, intimate portrait of the world as lived-in space.

Hong Kong-based artist and founder of the FAT (Free Art and Technology) Lab, a Web-based open source development lab, Evan Roth is well known for a variety of software hacks and apps that he has created. Roth's *White Glove Tracking* used crowdsourcing as a way of mapping data. On May 4, 2007, Roth asked Web-users to assist in isolating the white glove in all of the 10,000-plus frames of Michael Jackson's video, *Billy Jean*. Within 72 hours, 125,000 new versions of the data-tracked glove had been created. (You can access the highlights at http://whiteglovetracking.com/). The tracking reimagines the video with the glove as a focal point. What the glove does is a product of the ingenuity of each artist: rendered at an exaggerated size, set on fire, as a motion trajectory in space, as a slit-scan image, spatialized, as a cursor in motion, at variable speeds (the higher the glove, the faster the frame rate), etc. The *EyeWriter* interface is another of Roth's creations; it is an open source engine that he crowdsourced to allow a (or any) paralyzed graffiti artist to continue to make his art. Roth also created the first open source rap video for Jay-Z. He is the creator of GML, the graffiti mark-up language used for archiving and cataloguing urban tagging metadata. Roth's *Animated GIF Mashup* engine is neither so highbrow nor so altruistic as his other projects. Created in 2006 and released in a 2.0 version in 2010, it is an engine for producing original works using animated GIF (graphic interchange format) images. GIFs are the simplest kind of image on the Web. They are tiny and efficient, being composed of a maximum of 256 colors. GIFs are used mostly for clipart images, which are copyright-free graphics in cartoon form, for black and white line drawings, or single color images. Animated GIFs sandwich together a number of individual GIFs to create tiny, looped stop motion movies that are only a few kilabytes in size and only a few seconds long. They are the ultimate minimalist form.

Ani Up is a small video made in a workshop Roth ran using the *Animated GIF Mashup* machine at the Video Vortex 6 conference in March 2011. A high-profile advocate of free and open source materials, Roth drew these samples from online collections at Dump (http://dump.fm), Heathersanimations (http://heathersanimations.com), Gifsoup (https://www.gifsoup.com), and Tumblr (http://www.tumblr.com). The machine allows users to (singly or collaboratively) join animated GIFs together to create new, extremely lowbrow videos. Using the simplest and most crass form of image, Roth has devised an easy-to-use creation engine for designing copyright-free mashup videos. Designating only where the image is placed with a space for including background images and an audio track, the engine allows the user a fair amount of creative control over the overall look of the final piece. The *Ani Up* video, which Roth's workshop participants created, uses four basic kinds of images. First and foremost, there are cheesy clipart images. Second, cartoon characters and iconic figures (Wonder Woman running, Homer Simpson falling, Yoda breakdancing, Teletubbie playing an instrument, Elmo dancing, etc.). Third, viral media moments, which include singer Susan Boyle swishing her hips; an angry child fanning herself with cash; a slow-motion bullet making impact; a woman playing a stack of pancakes as if it were a beat juggler's turntable. And, finally, the largest group of images is comprised of recognizable people and personalities, including historical figures, celebrities, and actors. All of these figures are animated, of course, and often photoshopped as well. These include Joseph Stalin's headphoned head bobbing, one of Adolf Hitler's trademark gesturing speeches transformed into dance moves,

Marilyn Monroe blowing kisses, Rowan Atkinson gesticulating as Mr. Bean. Often these images appear repeatedly throughout the piece either simultaneously on the same screen or over the course of the video – as in the case of Mr. Bean where his visage is reproduced 15 times (once in an inverted position) in a series of side-by-side grids. Works do not get much lower brow than this, but the effect of putting all the pieces together with a soundtrack is to create a rhythmic piece where all the characters and objects seem to be gesturing or dancing in time to the music. The electronic soundtrack is Danny Drive Thru's "Ante Up," which itself is a mashup concerned mostly with drug culture and drug paraphernalia.

Another text that uses lowbrow images is published in the Electronic Literature Organization's second anthology (2009). Alan Bigelow's *Brainstrips* (2009) are reflexive remixes of copyright-free comics from the 1930s and 1940s that have been animated (or perhaps to say they have been rendered dynamic is a better description) in flash. Bigelow has rewritten their texts to explore philosophical, mathematical, and scientific issues. "What is Art?" shows an explorer with a pith helmet and a magnifying glass standing in front of a yellow background with a mathematical equation printed on it. Bigelow's altered text reads, "Finally, after centuries hidden from human eyes, the ancient formula is revealed." In the first frame, the scientist says, "Cadmium yellow … canvas width times Pi … equals artwork … minus artist's fees." He continues his musing in the second window, "As I've said all along: Art is simply a mathematical equation applied to different experimental contexts." In the third frame, he addresses another man with a lantern, indicating the yellow backdrop with its formulae, "With this formula, we can create masterpieces without using any artists." Clearly an ironic and self-reflexive commentary on the art of the remix itself, the text also parodies the scientific need to automate and classify things within a methodology. Just as the remix is alleged to be a mechanical act devoid of creative spirit, so Bigelow hilariously suggests that the formula of art could – in theory at least – make creative practice executable.

Bigelow's *Is Color Real?* explores the *Brainstrip* characters' self-awareness of the comic strip form and of 'their creator's' presence in their world. Frame one of four takes place "One day, someplace in a comic strip." A fisherman in the back of the boat says, "I sense a blackness all around us," referring to the black background around the comic bubbles. The captain says, "Uh, that's crazy talk," and then, in the next frame, whispers to the man in green, "(You think he knows?)" He refers to the man fishing and to his awareness that he is in a comic strip. The man in green responds to the captain, "(Maybe – your hand is breaking the frame.)" And indeed the Captain's left hand on the wheel has slipped out into the black border of the comic itself. Meanwhile, the sky

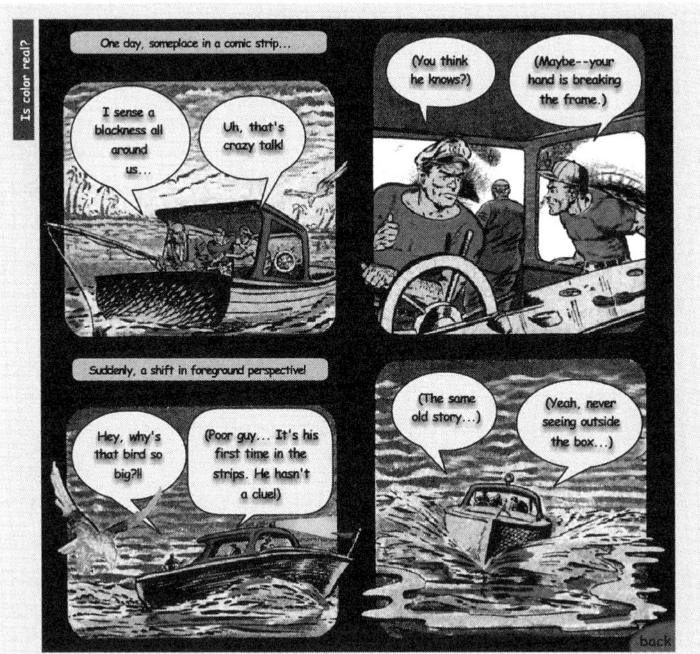

Figure 4 Alan Bigelow, *Brainstrip: Is Color Real?* (Original source: www.goldenagecomics.co.uk)

behind them is filled with the giant face of a woman (presumably the comic's artist) staring down at them, which they seem unaware of. The third frame is a medium shot of the boat with a gull in the foreground. As the drama intensifies, Bigelow sets the scene with a contextual statement: "Suddenly a shift in foreground perspective!" The man in the back of the boat calls out, "Hey, why's that bird so big?" The man in the cabin says, "(Poor guy ... It's his first time in the strips. He hasn't a clue!)" The fourth frame continues the conversation. The man in green says, "(The same old story ...) and the Captain whispers in response, "(Yeah, never seeing outside the box ...) as the water in their wake spills out over the black border. By commenting on their own medium, the comic form is deconstructed, and the characters' awareness of their own environment makes us look and look again at standard comic book conventions. Of course, the announcing of the changing techniques is a comical metatextual expression of a form's own conventions. By playing with our expectations for comics, the end result of Bigelow's exploration of this so-called lowbrow medium is to realize how sophisticated its rules and conventions actually are.

Another example of a reflexive remix is Australia artist Gabrielle De Vietri's movie, *Captcha* (2011). Her work is derived from Captcha, an online security device designed to thwart spambots from posting information on websites and blogs and other social media pages. Captcha, the utility, takes a distorted image of letters that are unreadable to the bots and asks humans to re-enter the 'word' or code as a security screening device. Using made-up combinations of letters is a further way of sorting out ever more sophisticated spambots from humans. In the manner of Lewis Carroll's *Jabberwocky*, De Vietri takes these nonce words and crafts a new world from them. She tells a dystopic story of strange creatures in a hostile land. In a five-minute voiceover narrated video that displays only keywords from the security program, it tells the story of Desmedowe and his or her people, the Redlemutes, who are en route to "Berbosel, the only cobbedes in the land to hold a cure for the dreadful enessemism that had riddled their bodies and minds since the water had been spiked with suplaespu." Fleeing the Sorsiders who enslave them, they push through the nestmettly brambles while trying to avoid alerting the predatory manackboar to their presence. Using their agoraphobic raglatts to find a suitable velerroc, they take refuge in the cave to escape a toxic rain of dowsfule, but instead they find themselves sandwiched between the expoustic stench of the advancing gob-spewing Sorsiders in their own filventhot and the burning downpour of cragrain. If the Redlemutes are detected and their conspeali frequency overheard, the Sorsiders will turn them into bendolowes and Desmedowe himself will be sentenced to labor in the rifinepit. The Soresiders' weapon, suplaespu, we are told, is secreted through hingschnott glands, which causes instant paralysis to the Redlemutes. A dangerous weapon to be sure. This is a first installment and we are left hanging in an enthralled state of suspense at the end as the suplaespu-spewing Sorsiders advance. The idea of remixing found words generated from a computer database's utility like Captcha for storytelling purposes makes for a beautiful, poetic meditation on hostile environments both in the narrative and in the poetic strains in the bones of the code on the walls of the digital world.

Sometimes corporations can create what seem to be reflexive remixes as an interactive marketing tool. Similar to Perry Bard's Twenty-first-century version of Dziga Vertov's *Man with a Movie Camera: The Global Remake, The Johnny Cash Project* invites viewers to submit their drawn interpretations of each frame of the music video of the last song Cash recorded in the studio. The video itself is a mashup of existing footage of Cash from documentaries and other materials, and fans contribute "a sea of one-of-a-kind portraits." The organic, ever-changing result is a stop motion version of "Ain't No Grave". Unlike Bard's mashup of Vertov's film though, which randomly changes the chosen clips for each scene every day, this Cash video has rankings for viewing based on different criteria voted on by users: highest rated frames, director curated frames, most brushstrokes per frame, most recent frames, random frames, pointillism frames, realistic frames, sketchy frames, and abstract frames. In addition, where part of the truly global nature of Bard's project is reflected is in how differently the project gets interpreted around the world on the basis of varying standards and users' abilities to access or fluency with technology, *The Johnny Cash Project* is severely restrictive, permitting only a single kind of illustration – created in the project's own interface – to be used. By contrast, Bard actively sought video clips from around the world. She found that no matter what she did, however, no African uploads were contributed to her project (except for one by a traveling Canadian). She produced her materials in multiple languages and had them distributed internationally. Assuming that users are already familiar with the singer's work, *The Johnny Cash Project* is aimed at fans; it is only in English and gives no background information on Cash. Bard led workshops worldwide to encourage entries from other cultural, linguistic, and ethnic points of view, but was constantly frustrated by the digital divide, which is alive and well. In Beijing, while the students were fascinated by her project – and even more astounded that an individual could conceive of such a thing – most had never uploaded a video before. They also did not know how to compress video files. They were thrilled to discover the ease with which they could achieve this through their cell phones, but had never before put their mobile phones to this particular use (Bard, 2011, 325). *The Johnny Cash Project* both solves this problem and considerably restricts the nature of the entries by providing a white-on-black background drawing interface that participants must use to create their image. This prevents the addition of remixed, photographic or digitally altered images from being submitted. It also means that there is a continuity in the final video that does not reflect the free nature of personal expression as is usually found in the age of networked communication (that we see, for example, in *White Glove Tracking*) – especially in a project that defined itself as a global enterprise. *The Johnny Cash Project* therefore ends up being a nonreflexive 'reflexive' remix.

Such crowdsourcing projects do bring to the fore the rising importance of the database in online communication and how it is changing our concept of video and audio-visual communication. In *The Language of New Media,* Lev Manovich dubbed the Russian filmmaker Vertov a pioneer of the database and database cinema. Database cinema is a new form. Something like *The Father Divine Project*, which bills itself as a database documentary, is a much simpler construct. It is a blog built from a database of historical footage of Father Divine's church and interviews given by Mother Divine in the 1990s that are arranged in such a way as to allow users to form their own opinions about Father Divine's Peace Mission Movement's religious practices and philosophies – presumably as he would have wished. The assembling of the

Father Divine narrative, however, is something that is done in the viewer's mind more than on or in the screen space of the work itself. Database cinema is generally something that is constructed on the fly from stored materials. What is most obvious in how the database is transforming cinema is apparent in Manovich's *Soft Cinema* installation, which foregrounds active spectatorship and collaborative processing as the viewer must put together the pieces of the work to make meaning. *Soft Cinema* has multiple frames of a single story that are generated randomly by a database and are spatialized on a single screen. Belgian-Australian artist Suzon Fuks has also created a work that explores those database-driven authorship issues in much more literal ways than Manovich's installation.

Fuks' *Waterwheel* is a reservoir and collaborative database on the subject and metaphor of water. The project spearheads three initiatives. It has an interactive website where people can upload, edit, use for research, or work on collaborative undertakings. A second facet is known as The Tap. This is a free webcam-enabled live performance and collaborate venue space. The site includes a whole palette of editing tools for group or individual live exploration. In The Tap there are two points of view: audience view and crew view (to enable real-time tweaking). This third part is called Fountains, and it comprises four spin-off initiatives generated by the initial project. These might be in any medium from tactical interventions, to performance, to books, exhibitions, community good works, to suitable spreadsheets. The fourth aspect is the watershed interface itself, which is a mandala-like archive where anyone can access all of the geolocatively mapped works. The site launches on August 22, 2011, so it will be interesting to see what works are generated by it. In both Bard and Fuks' works, the role and position of the artist has changed: the artist has become "a facilitator or mediator of the audience's interaction" with the work (Paul, 2008, 21–2). This is one of Bourriaud's preoccupations as well. He says,

> The tasks facing us today are to analyse how contemporary art addresses the viewer and to assess the quality of the audience relations it produces: the subject position that any work presupposes and the democratic notions it upholds, and how these are manifested in our experience of the work (quoted in Bishop, 2004, 73).

The democratic urge is now starting to break down some of the most conservative walls in the art world. In the past couple of years, highbrow institutions have started to jump on the remix bandwagon of this creative surge of energy. The result is festivals that are recognizing (and commodifying as 'art') the best digital works. It is lending remix practices a whole new air of respectability. Two of the most notable searches for great remixes have been run by the Guggenheim Museum in America, and Penguin Books in the U.K. (Other major festivals include the annual Video Vortex Conference in the Netherlands, Techcrunch France Remix, 24/7 DIY Video Summit, and the Streamy Awards – which, for the third year, are in financial trouble.)

The first YouTube Biennale, YouTube Play, was held in 2011. Sponsored by YouTube and the Guggenheim Foundation (comprising Solomon R. Guggenheim Museum in New York, Deutsche Guggenheim in Berlin, Guggenheim Bilbao in Spain, and the Peggy Guggenheim Collection in Venice), Nancy Spector, deputy director and chief curator of the Guggenheim Foundation, announced that "We are in a sense, inviting people to raise the standards of YouTube" (Artradarjournal.com). Not satisfied with the lowly YouTube aesthetic, the Guggenheim sought immediately to make it more respectable. Directly contradictory to YouTube's founding principles, this drive toward commodification even of the lowbrow is inevitable in consumer culture. Out of 23,000 entries, a shortlist of 200 was drawn up, and a panel of nine judges chose the 25 winners. Those 25 selections then received screenings at the four Guggenheim museums. Is this the American Idolization of YouTube? Spector saw the show as providing a new avenue of entry for those who would not otherwise have access to the art world. Again, they underline that "The jury is looking for innovative works that debate on, discuss, test, experiment with and *elevate* the video medium" (*Art Radar Asia*; emphasis added). One can only wonder how many lowbrow, postproduction *Ani Mashups* they rejected. The criterion was simple: each participant could submit one video entry of 10 minutes or less created within the past two years.

Because they disdain the YouTube aesthetic, most of the big fish they caught were professional artists of the caliber that the Guggenheim might have considered for the regular collection. Perry Bard's *Man with a Movie Camera: The Global Version* (that I have already mentioned), with its roots in the cinematic tradition, would be an expected highbrow pick. Australian photographer Kevin Loutit's stunning time-lapse photography called *Bathtub IV*, which combines wide-angle shots with a tight focus to create miniaturizing effects of great speed was also a clear finalist in the high art tradition. The 2009 Emmy Award-winning video for "new applications," Jerry Levitan and Josh Raskin's *I Met the Walrus*, was a predictable selectee for the Biennale, as the first Emmy award-winning Web-circulated film to be recognized with that award. This film was also shortlisted for an Academy Award in the animated short category. Using an interview that a then 14-year-old Levitan conducted with John Lennon when he snuck into Lennon's hotel room in 1969 as a soundtrack, Levitan worked with Raskin to animate Lennon's powerful visual language in a style reminiscent of old Beatles films. What the jury missed mostly were works that were conversations with other works, works that were in process rather than final products – like the remixes of *Downfall* (*Der Untergang*; Hirschbiegel, 2004) or works that are not per se 'original.' More surprising was the fact that some of the 25 finalists they selected did achieve excellence while still working with the parameters of more traditional YouTube aesthetics. These include Josh Bricker's *Post-Newtonianism*, Lindsay Scroggins' *Wonderland Mafia*, Christen Bach's *Bear Untitled: D.O. Edit*, Bryce Kretschmann's *Auspice: Sqomb*, and Martin Kohout's *Moonwalk*.

Bricker's *Post-Newtonianism*, which I discussed in Part Two, is a near-perfect mashup of American army Gulf War footage, the video game *Call of Duty 4: Modern Warfare* and a soundtrack that combines game footage with audio from a U.S. Military strike, a high-profile attack which killed two Reuters correspondents and a group of unarmed civilians. (A camera tripod was erroneously assumed to be an automatic weapon.) Another surprising choice is American sound editor Lindsey Scroggins' immaculate mashup, *Wonderland Mafia* (2008). She blends the idealized innocent animation of the hookah-smoking caterpillar from Disney's *Alice in Wonderland* with Three 6 Mafia's raw, rough, shocking gangsta rap music in perhaps the most uncomfortable 1 minute and 44 seconds you have ever heard. (In her director's video, she says that it took her between 200 and 300 hours to make the cuts and synchronize the sound.) As an assessment of hip-hop styled drug- and sex-laden themes and how uneasily they fall upon childish ears (as the world is so often revealed to the young), the work is disturbing in the extreme. The third YouTube-native work was *Bear Untitled: D.O. Edit* by the Berlin- and Copenhagen-based artist Christen Bach. Bach worked with a freeware program called GraphicsGale to create an 8-bit animation.[7] The simple charm of the 8-bit style is mixed with vibrant colors and what Bach calls an old school feel. Upending our expectations in the beginning as the man who ventures into bear territory with his shotgun, the story turns into a witty tale about a cross-species gay affair between a man and a bear. The monotone soundtrack and simple characters are brevity personified, and the man's suicide at being jilted happens so fast that the bear does not even have time to protest.

This Aborted Earth: The Quest Begins is by Boulder, Colorado filmmakers Michael Banowetz and Noah Sadona, co-founders of Untimely Films. It is a remix tale of the crusades in a Terri Gilliamesque style animation with Nineteenth-century line engravings. (I will discuss this work at length in the next chapter.) The most astonishing winners were Bryce Kretschmann's *Auspice: Sqomb* and Martin Kohout's *Moonwalk*. A dazzling example of the innovation that is possible with sampling, *Auspice* takes cuts of CNN News, overlays them, and slows them down so that they sound like a medieval choir.[8] More akin to net.art than viral YouTube, *Moonwalk* is a structural and sonic remix of the scroll bar as an ascending staircase of electronic music replayed like a dance step. Disposable art or brilliant digital rethinkings of existing material?

As I mentioned earlier, the Guggenheim has not been the only highbrow institution to jump on the lowbrow bandwagon. The ultra-literary publisher Penguin Books tried to channel digital culture through its six-author anthology featuring six short stories published over a six-week period in 2009. The *We Tell Stories 6* anthology was winner of the SXSW Interactive Best Experimental Award and *Best in Show Award* in 2009; it was also exhibited at the Museum of Modern Art (MOMA) as one of the works in the *Talk To Me Exhibition* in 2011. The six works are all remixes of recognizable literary works in the Penguin list, with a complement of social media to generate new effects. "The 21 Steps" by Scottish novelist Charles Cumming blends John Buchan's novel *The Thirty-Nine Steps* (1915), a celebrated and often filmed murder mystery, with locative media. Following a geographically based plot, "The 21 Steps" requires the user to navigate a Google map mashup in order to unravel the mystery. The story, however, is entirely linear, despite its geolocative structure, and is intended to be read sequentially through 21 'chapters.' Toby Litt's "Slice" is a retelling of *Haunted Dolls' House* by M.R. James through blogs and Twitter. During its six-day run, readers could receive and send email, blog postings, and tweets to stay involved minute by minute. Kevin Brooks' "Fairy Tales" remix Hans Christian Andersen's *Fairy Tales* along a choose-your-own-adventure model, where the reader inputs choices about variables in the story. The variables and names in the story may change, but it is always the same story. The children's book-writing duo Nicci French's "Your Place and Mine" is a retelling of Nineteenth-century French novelist Emile Zola's *Thérèse Raquin* love story in two voices in real time in one-hour installments every night for the six nights of the anthology's run. Matt Mason's "Hard Times" retells Charles Dickens' novel of the same name through statistics, and blocks pieces of information all in a downloadable pdf that is reminiscent of the spatialized, graphic images of Chris Ware. "The (Former) General" by Booker nominee Mohsin Hamid is a modern revisioning of *1001 Arabian Nights* in a spatial tale of exploration along the lines of old navigational computer games like Zork. A seventh tale, "Alice in Storyland" is Naomi Alderman's version and reinvention of Lewis Carroll's masterpiece, *Alice in Wonderland*. While all the works are elegantly written and beautifully realized as online texts by the company Six to Start, none of them do anything to further the field of digital narrative or break ground as born-digital works. Most are simply print stories that have been innovatively translated or masterfully adapted for the World Wide Web by this production company. Compare this to something really innovative like Stephen Fry's "MyFry" app for the iPhone, which creates a colorful mandala to be navigated in any number of ways, including themes and chronologies, and it makes plain just how unadventurous the Penguin anthology is. Ultimately, however, Penguin's works remain beautiful translations or adaptations rather than anti-narrative or multilinear works.

'Aberrant decoding' and atactical aesthetics

As digital remixing becomes the dominant mode of production in the Twenty-first century, the remixed work itself can be created using dramatically different techniques. There are seven distinct kinds of remixes or what I would call "atactical aesthetic" approaches. These are:

1 Sampling

2 Mashups

3 Remakes, adaptations, and/or intertexts

4 Capture, streamed content, and/or visualization

5 Surveillance art (which was discussed in the last section)

6 Archiving as an aesthetic form

7 Hacks.

These are the primary forms, but these methods are not mutually exclusive. A single work can mix and match two or more different techniques. It is on account of this fluid, promiscuous nature that Paul D. Miller, also known as Dj Spooky, calls remixing "cybernetic jazz," "rhythmic cinema," and "sonic sculptures" in his book *Rhythm Science*.

Sampling

Sampling, which came to fame first through its use as a musical technique, is the most familiar. Originally it came to the fore as a way to reuse often repeated cultural materials to create a danceable rhythm or beat. It pulls together or pulls apart elements to create an aesthetic – and now the principle is applied to video. Spooky aligns this with Joseph Beuys' theory of social sculpture, which he deems an act of collaging often recognizable audio visual tracks. According to Spooky, it is a way of giving structure to rhythmic forms across different contexts:

> Sampling is a new way of doing that's been with us for a long time: creating with found objects. [...] The mix breaks free of the old associations. New contexts form from the old. The script gets flipped. The languages evolve and learn to speak in new forms, new thoughts. The sound of thought becomes legible again at the edge of the new meanings. After all, you have to learn a new language. Take the idea and fold it in on itself. Think of it as laptop jazz, cybernetic jazz, nu-bop, ILLbient – a nameless, formless, shapeless concept given structure by the rhythms (Miller, 2004, 22).

"Sound and image divorce and reconfigure before they reunite in the mix," Spooky says (22). The creation of rhythmic structures in the sampling of audio-visual works is accomplished with tools of appropriation (including the use of found footage), and editing methods including "looping, repetition, erasure or compression" (Basilico, 2004, 30). Again, these tools are not new and many of them can be traced back to practices first conducted by the Surrealists. In 1936, for instance, famed American experimental filmmaker Joseph Cornell recut actress Rose Hobart's 1931 film *East of Borneo,* so that only some of Hobart's solo appearances on screen remained. The new film, called *Rose Hobart,* interlaced her footage with an assortment of clips from a scientific film about an eclipse, slowed the film speed to 16 frames per second (from the standard of 24 for talkies; silents had used a 16-frame standard), and reordered the original sequences. Over the top he added a dark blue filter,[1] and a new soundtrack – two looped tracks, "Forte Allegre" and "Belem Bayonne" – of the Nestor Amaral Orchestra's *Holiday in Brazil*, a Samba album Cornell found in a junk shop.[2] (At the 1969 screening, Cornell merely played the record during the event, and so its likely that he did the same at the original screening.) The 19-minute resulting film is hypnotic and dream-like "in which the characters appear to 'move with a peculiar, lugubrious lassitude, as if mired deep in a dream'" (Yeo, 2004, 16). A piece out of time, this is now seen as a transformative work in the history of avant-garde film. We might also consider *Rose Hobart* (a respected actor of the 1930s and 1940s who bears an uncanny resemblance to Kim Novak) as the first fanvid.[3]

Another gallery-projected film that follows the formula of elimination as a creative practice is South African artist Candice Breitz's *Soliloquy Trilogy* (2000). Using the audiotrack as her guide, Breitz reduces three films – Clint Eastwood's *Dirty Harry* (1971; 6:57), Sharon Stone's *Basic Instinct* (1992; 7:08), and Jack Nicholson's *Witches of Eastwick* (1987; 14:06) – to the leads' soliloquies. Projected in three channels, the elimination of all other content except the central, isolated figure, with whom the camera demonstrates a near-lurid fascination, results in a tunnelizing effect as we resee and rehear and "stalk" these characters (Beccaria and Breitz, 2005, 43). "[C]ut and pasted" as they are into their isolation, with their own speeches bumping up against each other in the absence of any other content, seems to reduce them to the level of pure psychoses:

> As the *Soliloquies* devolve into narrative babble (Sharon), taciturn repetition (Clint) or manic tyrade (Jack), one is reminded of the extent to which language depends on context and community for its meaning. [...] Questions are asked but never answered, comments are left without response (Beccaria and Breitz, 2005, 43).

While other characters' faces, voices, and reactions are blackened out, the star's voice remains, uninterrupted and unanswerable. The assemblage of the remaining parts results in entirely new movies. For instance, the sexual tension in *Basic Instinct* is lost in the new version and "Stone's character rambles through a series of repetitive, narcissistic, and apparently unmoti-vated statements" (Beccaria and Breitz, 2005, 24). Eastwood's Inspector and Nicholson's Devil similarly lose their situation in space and time and become models of neuroses instead. Through elimination, motives and actions are transformed. Breitz is a child of the video revolution, not the digital one, but her

> creative process is defined as a process of selection and translation. Her artistic strategies reflect the 'cut-and-paste' logic that has spread so dramatically with the increasing availability of digital technology, a logic that we now find everywhere – in film, video, music, photography and fiction. The meaning of her work is ultimately linked with the formal strat-egies that she chooses" (Beccaria and Breitz, 2005, 20–1).

Beccaria goes on to say that this process not only connects her to her digitally remixing contemporaries, but that "Breitz's use of 'cut-and-paste' also harks back to the twentieth-century collage and montage" where her work serves "not only to open up the picture plane and let in the chaos of every day life but also to challenge the hierarchy of 'high' and 'low' culture" (Beccaria and Breitz, 2005, 21). What her work ultimately reveals, according to Breitz's curator Marcella Beccaria, is "Breitz's conviction that Hollywood cinema, like other forms of mass culture, is designed for passive consumption, providing neither stimulus nor any possibility of access to its monolithic structure" (Beccaria and Breitz, 2005, 24).

Sampled digital art enables precisely this ability to talk back to the monolith of mass culture and to drive wedges in the surface of its seemingly unbroken and unbreakable structure. Erasure, looping, repetition, compression, and elongation as editing techniques of found footage have the effect of shifting the emphasis back on to the effect rather than on the content. As a metatextual gesture, the act of sampling becomes a way of critiquing media spectacle, consumer culture, and digital technologies themselves. By introducing rhythmic elements as well (often simulating a beatbox effect), sampled remixes can underline particular aspects of a work. Among the most celebrated DJs in the world are a British group who go by the handle of Addictive TV. Their classic *Star Trek* sample "Beam Up The Bass" may be one of the finest remixes ever. Using the trademark weird science effects of the series, including communicators, phasers, and transporters as musical beats, and pairing them with some of the worst scenes from the series, they create a wickedly funny commentary on media and communications themselves. By toying with the sounds and protocols of ship communications, they explore telephony, noise, and broadcast. They also render James T. Kirk's preparations for his fight scene with the Zork as a percussive musical interlude revolving around much repetitive stick banging. Stutters and idiotically bad lines get repeated over and over as rhythmic stammers.

Another extremely talented group is the duo Dan and Dominique Angeloro, Australians known as Soda_Jerk. Their 52-minute video *Pixel Pirate 2* has been featured both at the Sydney Underground Film Festival and at the Electrofringe in Townsville, Australia. In *Pixel Pirate 2* Soda_Jerk, in collaboration with Sam Smith, use sampling as a way of critiquing piracy and copyright. Blending more than 300 samples from science fiction, action, and biblical epic films and other sources, *Pixel Pirate 2* is a virtuoso performance of Afro-Futurism, and anti-copyright practices and beliefs. Moses as played by Charleton Heston is cast as the master copyright controller and arch-enemy of a troupe of lunar video pirates who are trying to liberate creative practice from the Copyright Commandments. The film's superhero who fights for samplers' practices is a cloned and temporally relocated Elvis Presley, who, when needed, transforms into a very green Hulk. Defeating Jedi knights,

Lara Croft, and even the Ghostbusters, Elvis-Hulk's job is to assassinate Moses and alter the trajectory of the future of video. Much of the film's continuity derives from the use of a red 1966 Thunderbird footage that has been a favorite in films from Elvis movies to *Thelma and Louise* to Japanese B science fiction films. As the Thunderbird is remixed from multiple films it threads together many different segments of found footage. At the same time they posit turntablism as a language of the future, a language more advanced and universal than our own. At their website, they describe themselves as using "audiovisual sampling to create speculative narratives that review and interrogate historical events and cultural trajectories" (http://www.sodajerk. com.au/info.php) through video installations, performance lectures, and the underground film circuit.

> By atomising and reassembling recorded culture they aim to manufacture counter-mythologies of the past that open new possibilities for thinking the present. This conceptual strategy also informs their practical methodology where sampling becomes a means of synthesizing space-times to create alternate historical realities, producing a form of radical historiography that merges research, documentary and speculative fiction (http://www. sodajerk.com.au/info.php).

By rethinking the whole process of creation through copied materials, Soda_Jerk produces witty and seamless multilayered sampled videos with the labor-intensive process of frame-by-frame image masking. Like Girl Talk's elaborate composition method, Soda_Jerk leaves little of the original works untransformed by the creative process. Unlike Girl Talk's performances or Addictive TV's vids, these fast-paced works do not produce anything that is danceable, but they are samples all the same. Always in the foreground in this illegal film is the problem of copyright. None of the samples have been cleared – and once a videotape replaces Kubrik's staid monolith on the moon or Elvis' face is superimposed over Heston-as-Moses' – they feel it is not likely that the corporations would be willing to allow this sort of manipulation (skynoise). Pioneers of the feature-length remix, this duo is reinventing remix on a scale not imagined since Pixar made *Toy Story*, the first computer animated film, in 1995.

Christine Paul has observed that "The work of art in the age of digital reproduction ... takes instant copying, without degradation of quality from the original, for granted" (Paul, 2008, 28). This is plainly evident in Omar Fast's compelling, paranoid digital video vision, *CNN Concatenated* (2002). Using a database of 10,000 words, breaths, and pauses, Fast assembles a psychotic imperative where CNN pundits and anchors deliver a series or sequence of paranoid messages and instructions that will ultimately lead us to suicidal,

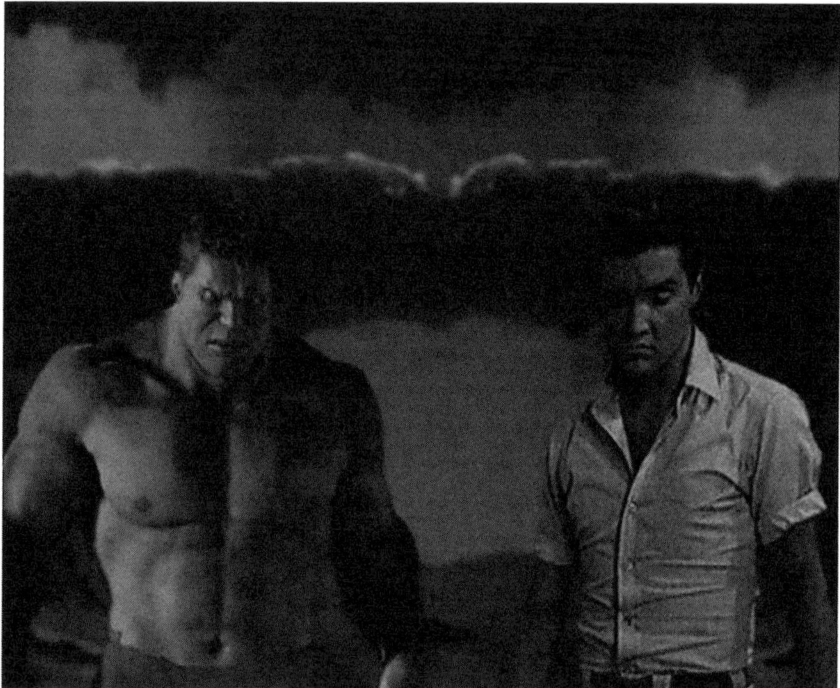

Figure 5 Soda_Jerk, *Pixel Pirate 2* (2011)

violent, or murderous acts. The media deliver a constant barrage of negative messages, about the listener's body image, for instance, and of anxiety-fostering bad news. Fast turns that into a series of messages from the set with each speaker uttering only a single word or breath:

<<This is CNN>> Listen to me. <<breath>> There's a few more things that you need to hear. <<breath>> Don't talk. <<breath>> Don't move. <<breath>> Don't even react. <<breath>> Actually, don't do anything at all. <<breath>> Just get near me already, you hypocritical opportunist, fake, phony, con artist, sellout, lip-serving, limousine liberal, white chicken shit mother fucker. <<breath>> What's the matter? <<breath>> Have I hurt your feelings already? <<breath>> Can't you speak? <<breath>> Can't you say anything? <<breath>> Have you lost your voice all of a sudden? <<breath>> Maybe you never had anything to say to begin with. <<breath>> Has that occurred to you? <<breath>> Well. <<breath>> Let me tell you something. <<breath>> You are shallow and weak. <<breath>> You are constantly criticizing everything, but the truth is you have never produced anything of enduring significance, and now you're finding out just how inconsequential your opinions have been all along.

The power of the media becomes the power of psychosis and it speaks to all our innermost fears, desires, and compulsions.[4]

Using black and white line engravings from the Nineteenth and early Twentieth century as raw material for sampling, Michael Banowetz and Noah Sadona's *This Aborted Earth: The Quest Begins* (2010) crafts a hybrid tale of a hypothetical crusade that spans the 'failures' of the Renaissance, the Enlightenment, and the Industrial Revolution. "Sodomites of the Renaissance have been eviscerated," "liberals of the Enlightenment obliterated," "pinnacles of philosophy decapitated," and "friends of terrorist infidels annihiliated." As a sample that is also a mashup, this text blends assorted histories and times to create a horrific alternate world rooted in an apocalyptic vision generated by past holocausts. Using witty black comedy in the style of Terry Gilliam's *Monty Python* animations, *This Aborted Earth* (of which this is only a first installment) features a ferociously righteous knight, General Thaddeus Twer, and his trusty steed, a canonized sarcastic horse (who was flayed alive) by the name of Saint Gluestick. Twer and his horse carry out their bloody mission to find the planet's savior, "the last good man." Twer is particularly bloodthirsty, having slain more than 7000 nonbelievers, pagans, and heretics. He is what Pope Pious XXXIII calls "a true patriot." He dubs himself "the Fist of God, Sword of the Lord, Obliterator of the Infidels." Charged with the Pope's holy mission to fulfill God's decree, they set out with more zeal than is really necessary to cleanse the Earth in their quest.

Dan Warren's *Son of Strelka, Son of God* is really remarkable for taking sampling to a whole new level. Struck by the rhythms, the syntax, and mythic language used by Barack Obama in the audiobook version of his autobiography *Dreams from my Father: A Story of Race and Inheritance*, Warren sampled Obama's language to foster an entirely new and distinctly unique creation myth. Warren describes his laborious creative process:

> Bit by bit I've dissected Obama's self-read autobiography into thousands of very short phrases, usually one to ten words or so, and have used these snippets to tell a completely different story from the original. I've then set the story to music. [...] Broadly speaking, it tells the story of an ugly dog-faced demigod who recreates the world after it is destroyed. It's about thirty minutes long, and lies in some weird grey area between audiobook and electronic music (forums.somethingawful.com).

Son of Strelka,[5] *Son of God* starts with the story of a minor god who is born from a fruit tree and who speaks with all life that roams the Earth until he discovers within himself the power to conjure up living things and beings with his words. His yellow dog of a son, Stanley Steamer – who narrates the story from the perspective of a post-apocalyptic world – has his father's power but does not embrace it. First, he goes on a quest to find out his father's motives for destroying the world. He tours the devastation of the world and consults spiritual guides – the Turtle of Hindu mythology and Buddha – to find out his own nature and to assess whether he should shoulder the burden of his birthright. He must acquire enough faith in himself to be able to exercise his own power without abusing it.

The story has a strange, haunting rhythm more akin to music or oral storytelling than to narrative. As a narrative, the work is abstract in the extreme, despite being told in concrete, visual language. Part of its strangeness arises from its oddball, sampled grammatical structure – part of it is Obama's trademark style of oratory. The end result is many sentences that are verbless fragments. Sometimes the syntax is hard to follow because Obama's emphasis on particular words or sentences does not fall where we would expect it to due to the original being transposed. Part of the story's charm is the unexpected nature of the language and its juxtapositions, but those surprising elements sometimes distract us from the plotless journey; in fact, this work probably bears more resemblance to a modular oral tale like Homer's *Odyssey* than it does to the style of a print narrative. Other sentences are lists constructed as run-on sentences with a large number of phrases joined together. A couple of examples will help clarify. After his father's genesis, the cities grow upward, and people hunger for personal prosperity. Their needs change: "People began to spend more time inside. They bought a house and found themselves bridge partners" ("The Golden Age"). The incongruity of 'bridge partners' derails the trajectory of this sequence, becoming the most memorable thing, but leaving us uncertain as to what kind of a marker of civilization this actually is. In another such sequence, we see that Stanley's gift is also linguistic one. Unlike his father, it is ultimately his abilities as a storyteller that give him the power and imagination to create a new and better world. When Stanley conjures up a new and better vision, he quickly discovers that, while at first everyone is grateful, his creation is far more complex than he anticipated:

The beauty, the filth, the noise, the excess, I discovered that I couldn't escape it if I tried. Cars, motorcycles, bicycles, rickshaws, buses, and [indistinquishable] filled twice to capacity. The square-jawed men in fast cars, the Marxist professors and structural feminists and punk-rock performance poets, all unsmiling men in suits and ties and mud-clogged groupies, slender models in fashion magazines, all jostling to be heard ("Restoration").

Imagined as a list, the world thereby become a big additive, complicated, twitchy pile constructed in language. Another aspect that makes the story hard to follow is the initial animated sequence. Only the first two parts have been animated so far; the first by Ainsley Seago and the second by Adam Bozarth. While the second follows the narration fairly straightforwardly, the first confuses the speaker with his father. Drawing the fruit that is to become Strelka as the narrator in the first video of the work muddies much of the story that comes later, leaving the reader wondering who exactly the narrator is and whose tale he is telling. Some of that confusion likely would have existed in any event, since the narrator does not introduce himself until "The Fall," the fifth of nine parts. The story's sections or movements are called "The Creation," "The Golden Age," "Son of Strelka," "The Decline," "The Fall," "The Turtle," "The Buddha," "Restoration," and "Son of God."

Warren's sampling process was a complex as world creation. Working with 2900 audio phrases, he was particularly interested in odd syntactical combinations and unusual phrasing. Working in Acid software to order his samples, he constructed the sentences one by one:

> Acid lets you see the file browser and the track layout at once and I'd named the files with the text they contained, I could bring in audio clips in much the same way that you do with those refrigerator magnets that have little words and phrases on them that you assemble into a poem. There was a lot of trial and error involved in making it sound as natural as I could – i.e., frequently Obama would say the right thing, but in the wrong tone of voice. Sometimes I would run into a spot where I really needed him to say something that I didn't have a sample of, and then I had to go back and either do some very targeted sampling (here I benefitted from having a searchable ebook of it as well) or I had to do some very careful editing to patch phrases together from multiple sources in the book (Warren (as Mofolotapo)).

The nine-part project is about half an hour long and took years for Warren to complete. He estimates that hundreds of hours went into each stage of the sampling, assembly, composition, and recording of sound effects. It cannot be sold, of course, since it would violate copyright law. You can watch it in video form on YouTube for the first two chapters and listen to the free audio download as a whole. As a work of art, it is a testament to the hope that a new administration engendered in a person in a particular time and place, and is surely the most unusual incidence of sampling yet seen. Furthermore, in the gift economy of the Web, this sample is part of a chain letter as those who are captivated by the narrative can give it a new shape in their own eye as an animation for each part – and there is no limit to how many different versions of each section might be created, like for Bard's *Man with a Movie Camera*. As Bourriaud (2002a, 13) notes,

> With music derived from sampling, the sample no longer represents anything more than a salient point in a shifting cartography. It is caught in a chain, and its meaning depends in part on its position in this chain. In an online chat room, a message takes on value the moment it is repeated and commented on by someone else.

On the Web, this can become a method of crowdsourcing a creative project.

Mashups

A mashup joins two or more cultural objects to make a new object, as in Danger Mouse's *Grey Album*; it marries The Beatles' *White Album* with Jay-Z's *Black Album*. These are often diametrically opposed pieces or works of radically different periods or styles – as with Seth James' *Pride and Prejudice and Zombies,* a mashup of Jane Austen's tale of regency manners and 19th-century romance, *Pride and Prejudice*, with the conventions of the zombie genre. A mashup is clearly transformative, creating something unique from the originals while retaining the originals in a recognizable form. It is important to note too that mashups are not just a creative practice but a fundamental principle of Web design. In designing pages there are several different kinds of mashups that may be applied: *aggregators*, which combine feeds and information from different Web-based sources like Google reader (for RSS – really simply syndication – subscriptions) or popurls (a customizable news aggregator); *search functions* like CloudMe (which unites Yahoo, Flickr, and YouTube) or programmableweb (which searches for 2.0 sites), zontube (which joins LastFM, Amazon, and YouTube for music searches); *visualizers*, which render data graphically or in visual patterns, as in Coverpop (1001 top-rated YouTube videos updated twice weekly); *geographic mashups*, which take data from other sources and plot them on a map as in Earth Album (Flickr + Google Maps), mashplanet (a geographic search engine), or Twittervision3D (which displays tweets around the world in real time); *mobile mashups*, which converts feeds into a portable format, as in FruCall (comparison shopping); *games*, where players guess information about particular objects, like PhotoMunchers, Wordhunter Xtreme (a mashup of Urban Dictionary and WordNet), or flicktionary. In addition, mashups usually have a *visual aspect*, which visualizes data, and a *technical aspect*, which joins different kinds of feeds. Increasingly artists are using this aspect of digital technology to perform cultural criticism and to queer (in many senses of the word) the mainstream.

As Christine Paul notes, "Recontextualization through appropriation or collaging, as well as the relationship between copy and original, are … prominent features in the digital media" (Paul, 2008, 27–8), but the extra and perhaps irresistible aspect of digital mashups is that they enable professional quality remixes of work that look as if they were actually made by Hollywood studios or by mass-media producers themselves. Using digital materials is a prime opportunity for *détourning* – hijacking, if you will – or subverting the kinds of visual arguments spectacular culture excels at in a manner that would not be possible in written or spoken language on their own. As Guy Debord says in his prescription for "Methods of *Détournement*": "one is not limited to correcting a work or to integrating diverse fragments of out-of-date works into a new one; one can also alter the meaning of those fragments in any appropriate way, leaving the imbeciles to their slavish preservation of 'citations'." He advocates that:

> Any elements, no matter where they are taken from, can serve in making new combinations. The discoveries of modern poetry regarding the analogical structure of images demonstrate that when two objects are brought together, no matter how far apart their original contexts may be, a relationship is always formed. Restricting oneself to a personal arrangement of words is mere convention. The mutual interference of two worlds of feeling, or the bringing together of two independent expressions, supersedes the original elements and produces a synthetic organization of greater efficacy. Anything can be used (Debord, 1956).

It is in building a bridge across the gap in the original highbrow or lowbrow or both kinds of material and in the politics made apparent in the process of recontextualization that we are invited into the act of "aberrant decoding." A semiological term coined by Umberto Eco, "'aberrant decoding' refers to situations in which the receiver of a message fails (intentionally or not) to interpret the message according to the set rules governing signification ('codes') used by the sender of the message" (Wees, 2002, 16). It is precisely in such deliberate misreadings that the mashup revels as a creative practice. This is not a new technique, but until digital technologies democraticized film production and editing, making them accessible to all, that 'disruption' (Wees' word), or what we might call active readings of mainstream images, was profoundly difficult. Mainstream cinema is at its core a process of normalization of stereotypes and a vehicle for upholding the status quo. Art cinema practice is its *doppelgänger*. It seeks always to undermine and trouble this "artificially imposed order" (Wees, 2002, 5) of class-enforcing, gendered and racialized images:

> The question is, what kind of "disruption" is possible when the influence of experimental/avant-garde film is so limited and the distance between the two realms of filmmaking is so great? Indeed, according to conventional wisdom, Hollywood and the avant-garde occupy opposite ends of the cinematic spectrum: the former makes art for a mass audience, the latter makes art for an educated elite.

Or do they? Those categories seem to be collapsing as everything becomes digital and networked. As digital technologies elide that distinction more and more, mashups reflect destabilizations, especially when it comes to gender roles. As I have already explored in relation to Cornell's *Rose Hobart* and Breitz's *Soliloquies*, the art practice of the experimental tradition of found-footage films has much in common with the *détournes* and aberrant decodings of lowbrow remixes within digital culture – and it is particularly well suited to resituating or 'queering' the original films and contexts. In several ways, the democraticization of digital technologies has made experimental techniques accessible through the rise of the mashup. And reading against the grain or queering the text has long been a highbrow art practice. As an interrogation of gender politics and sex roles, the mashup is a tool that is perfectly suited to critique and appropriate to the formalist style of Hollywood melodrama and sensationalized epics that it prefers.

One of the most celebrated art house mashups is Chilean-born Spanish filmmaker Cecilia Barriga's *Meeting of Two Queens: Encuentro Entre dos Reinas* (1991). Barriga's video enacted an onscreen realization of Marlene Dietrich and Greta Garbo's (reputed) real-life love affair. Long a cult favorite on the gay art and video circuit, this witty romp joins the two actresses onscreen – in many of their 1930s roles and period costumes – through the grammar and syntax of classic Hollywood film: longing looks, meaningful glances, and shared cigarettes. Bringing the two lesbian (or bisexual) stars on screen together with masking and dissolves, Barriga connects their faces, looks, and gestures. Wees (2002, 9) argues that Barriga's:

> most effective technique ... is intercutting shots that make it appear as if the two women are responding to each other without actually appearing on the screen at the same time. Due entirely to Barriga's montage, they exchange looks, talk to each other (over the telephone and face to face),

and meet in various locations – a hospital, a library, the Russian Imperial court, and a bedroom, where, through eye-line matches, Garbo watches Dietrich undress and impulsively begins removing her own clothes. By the end, the lovers appear to have parted. Each is seen embracing a man, and a tear containing an image of Garbo's face slides down Dietrich's cheek just before the end.

By reusing their signature scenes dislocated from their tragic (or comic) trajectories – *Queen Christina* meeting the *Scarlet Empress* or the *Blonde Venus* encountering *Anna Karenina* – Barriga resituates the actresses in a different narrative of the so-called love-that-dare-not-speak-its-name that we never saw onscreen. A sort of writing beyond the ending, Barriga gives us the possibility that the characters' and stars' destinies might not be so unchangeable after all. In a similar vein a year later, artist-filmmaker Mark Rappaport undertook a project called *Rock Hudson's Home Movies* (1992). He spliced film clips of Rock Hudson with a masked in onscreen hunk of an actor-narrator (Eric Farr) who is a Hudson look-alike. Farr provides a commentary on gender relations in Hudson's homosexual lifestyle versus his on-the-surface (at least) straight portrayal of hetero-normativity in films. While we can read these works as mashups from a Twenty-first-century standpoint, there are significant differences. Barriga's 14-minute and Rappaport's feature-length videos are far more coherent and sustained undertakings than most digital mashups, for one thing. They also present footage that is considerably altered as a part of its recontextualization. The queering of the mainstream in mashups under digital culture in a time that is considerably freer in terms of its gay practices and politics has no concern with autobiographical, biographical, or authorial/directorial intentionality as did Barriga and Rappaport. Instead, queer subtexts are a given and become starting points for conversation, statements of political allegiance, the insertion of sex into the homosocial bonds of friendship, and wish-fulfillment for real, serious gay visibility (as opposed to the more usual comic figures or objects of derision) in the mainstream.

A well-known example of a gay mashup is the queering of former President George W. Bush and former British Prime Minister Tony Blair's solidarity and political allegiance in the wake of the September 11 attacks and subsequent invasion of Iraq. "Bush and Blair's Endless Love" was originally created by Johan Söderberg for broadcast on Swedish television on the *Kobra* program, and later went viral on the Web. Blair and Bush's political 'marriage,' as they launched a united front in the Iraqi War, is transformed and translated via a mashup of sampled materials depicting them singing a duet of Diana Ross and Lionel Ritchie's saccharin love ballad.

More often, memes that go viral lend themselves to repeated and multiple interpretations. One such queer number was the trailer for the gay cowboy love story *Brokeback Mountain* (2005) by Taiwanese-American director Ang Lee. Generating hundreds of versions, the mashups take the formula of the trailer for Lee's painful and tragic tale of forbidden love among cowboys (based on a short story by Annie Proulx, and a screenplay by Larry McMurtry and Diana Ossana) and then apply that template to other mainstream, straight films. Through exploring irresistible urges, shame, and deep feelings, the meme allows for the queering of virtually *any* text that includes a close relationship between two men. The trailer has undoubtedly been seen many times more than the film itself.

While, on the one hand, the creation of one of these versions is an exercise in one's ability to create a professional-looking trailer, on the other it is a rhetorical strategy for creating while working within strictly defined parameters. The mashups use Gustavo Santaolalla's heartfelt Oscar-award-winning guitar music "The Wings," and adds variations on the original trailer's titles, which read:

There was a friendship
That became a secret
There are places that we can't return
There are lies we have to tell
There are truths we can't deny
"I wish I knew how to quit you."

The mashups then use original elements specific to each new film and recontextualize them within this template. This usually includes blatant examples of male sexuality – such as bare chests, aerobic physical activities of a non-sexual nature, and any overwrought expression of an emotional nature – that took place in the original. While many of the resulting trailers are parodies on manly films and series as wide-ranging as *Top Gun, Spongebob Squarepants*, and *Back To The Future* that get played purely for laughs, the most successful 'bromances' are the ones that find the same formulaic story of the blossoming of a friendship between a more experienced man and a more vulnerable one in the original. How the pair then suppress their feelings and keep their love a secret – especially from the women in their lives (who resent their emotional unavailability) – is what plays out in each trailer.

This is the essence of the *détourne.* Debord says, "Such parodical methods have often been used to obtain comical effects. But such humor is the result of contradictions within a condition whose existence is taken for granted" (Debord, 1956). The most successful of these mashups are not parodic at all in fact, but as deadly serious as the original film. In addition, the more familiar one is with the originals, the more satisfying they are. Ironically the movie studio lets these pirated videos remain online because they are seen as parody rather than as currency for a manifestation of and vehicle for queer-speaking culture. This meme seems repeatedly to come to rest most comfortably with the fantasy and science fiction genres, perhaps because of their heightened drama and tendency toward melodrama. *Brokeback of the Ring* by onetrickdog explores the homosocial world of J.R.R. Tolkien's *Lord of the Rings* trilogy (Jackson, 2001, 2002, 2003) through the love and friendship of two hobbits, Frodo Baggins (Elijah Wood), who bears the burden of the quest, and his protector Samwise Ganges (Sean Astin). *Brokeback Batman* by ohyeahvideos queers the teacher–child bond between Bruce Wayne (Christian Bale) and his 'lonely' bachelor butler, Alfred Pennyworth (played by Sir Michael Caine) as they share the "dark secret" of Wayne's superhero activities in Christopher Nolan's 2005 *Batman Begins*.

Broke Trek by Zebona has had more than a half a million views. The friendship between the Captain, James T. Kirk (William Shatner), and his human-Vulcan first officer, Spock (Leonard Nimoy), has long been a subject for erotic and queer fan fiction. This trailer translates and reinserts the *Brokeback* formula into the original *Star Trek* series so that it reads: "It was a five year mission … /To seek out new life and new civilizations…/They didn't think the one thing they'd find was … love!" The mashup uses excerpts from episodes where the normally emotionless and stoic Spock feels emotions – that are usually imposed on him or acquired through outside alien interventions, malfunctions, hormonal cycles, or intoxication by foreign bodies – and applies them to the affection the two friends have for each other in the series to craft a blossoming love affair. The *coup de grâce* for Zebona is in situating moments of emotional excess (as, for instance, when Spock is leaning over a possessed Kirk in a hospital bed) as transcoded material for sexual activity.

Two *Star Wars* versions, which have had more than a million hits, work in a similar way. *Star Wars: The Empire Brokeback* by SLC17 explores the comic relationships between two androids, C-3P0 and R2D2, and their mechanical malfunctions. *Star Wars: The Empire Strikes Brokeback* by mfish4 is the tragic love story of the friendship between Anakin Skywalker (Hayden Christensen; who will come to be known as Darth Vader in later films) and his teacher, Obi-Wan Kenobi (played by Ewan MacGregor). In the former, the mashup uses the role of the droids as outcast and obsolete equipment combined with what seems to be "technological malfunction" (in the original it was actually a message implanted in R2D2) to explore gay themes and humor. Male connectors become phallic objects, bomb drops and an exploding deathstar become recoded as expressions of male orgasm. The droids' story seamlessly blends the romance and fairy-tale aspect of "In a galaxy far, far away" with "There was a friendship/That became a rebel alliance." The jedi 'force' in the Star Wars universe becomes "a force that penetrates us." Droids are discriminated against, denied entry to what is recoded as a straight bar: "We don't serve your kind here." While spying on Darth Vadar's ship, R2D2's inexplicable activities give the rebel's presence away; in the mashup, this is translated into secretive sexual behavior. Perhaps the most brilliant stroke is in the translation of R2D2's binary speech (which only C-3P0 can understand in the original) into subtitles of the most often quoted lines from *Brokeback Mountain.* In response to a query from C-3P0, "Where do you think you're going?" R2D2 responds with the *Brokeback* line "That's nobody's business but ours" and later in the trailer R2D2 tells C-3P0, "I don't know how to quit you."

Star Wars: The Empire Strikes Brokeback by mfishe4 may be the most lovingly crafted *Brokeback Mountain* meme trailer of all. Nearly 3 minutes long (most are 1:30 to 2 minutes long), it explores the relationship between padawan Anakin and his teacher Obi-Wan. Obi's heartbreak over Anakin's evil decline as he, first, becomes more and more unruly, and then ultimately embraces the dark side outright, is translated into heartbreak over the shame at their passionate impulses. The *Brokeback Mountain* taglines become: "It was an apprenticeship/That became a secret/There are places we can't return." 'The lies they have to tell' harkens back to the chronologically later trilogy of movies (that were made first) where Obi-Wan (played by Sir Alec Guinness) who is now Anakin's son's, that is, Luke Skywalker's (Mark Hamill's), teacher, tells Luke that his father was killed by Darth Vadar – instead of telling him that his father *is* Darth Vadar. The truth that cannot be denied becomes Obi-Wan's agonized wail when Anakin lies defeated and apparently dying after their battle: "You were my brother, Anakin. I loved you."

The studios allow these trailers to remain on YouTube in part no doubt because they are free advertising, but also because they are classified as parody and so, therefore, permissible under current copyright law. Their genuine expression as queer culture, however, just goes to underline the fact that one (wo)man's love story is another's humor. As William C. Wees has noted in "The Ambiguous Aura of Hollywood Stars in Avant-Garde Found-Footage Films," "found-footage films nearly always have the effect of bracketing the images and calling attention to them *as* images, as constructed representations, and therefore as something that can be constructed or 'undone'" (Wees, 2002, 4). Alien cultures resonate particularly deeply for gay artists and citizens who have historically grown up feeling othered and outside of mainstream culture. Science fiction has long been a space into which alternate lifestyles could be inserted – as in Canadian filmmaker Dara Gellman's three-minute 1998 video installation, *Alien Kisses.* Her video takes a gay kiss from the 1990s television series *Star Trek: Deep Space 9.* One of the first lesbian kisses depicted on prime-time television, it was a gay scene that was sanitized and rendered permissible because the characters are aliens with a complicated life cycle. The Trill species is a symbiotic one, where a parasite inhabits a human host. The parasites – known as elders – far outlive the bodies of their hosts, who can be of any gender. The episode, which is called "Rejoined" (1995; Season 4), explores the Trill's taboos of having contact with people from the host's past lives; in the present time of the episode, the former heterosexual couple now both have female hosts: Jadzia Dax (Terry Farrell) and Lenara (Susanna Thompson). The scene is prefaced in such a way that the viewer is to understand they are not really queer, just spouses separated by insurmountable barriers. Gellman takes the original clip and translates this kiss into an expression of gay love as something that is alien and apart, but no less erotic for being so. Gellman also transforms the video footage to make it even more alien. She slows the playback speed, and adds noise and a ghostly blue tint. Queerness thereby becomes unearthly. Similarly, filmmaker Canadian Kevin Kelly's video *A Supernatural Premiere* (1995) is a mashup of 1960s science fiction, supernatural, thriller, and just plain weird footage (including scenes from *Lost in Space,* Hitchcock's *The Birds,* and *I Dream of Jeannie*), and his narrated tale of a physical encounter that engendered his newfound awareness of being gay at the age of 7 is rendered alien.

While the science fictional universe has long been known for exploring issues outside of the sexual normative parameters, 19th- and early 20th-century fiction has also often opened space for and awareness of romanticized female friendships that were not overtly sexual in the originals. Montréal performance and video artist and "Queer Revisionary" (as T.L. Cowan dubs her) Dayna McLeod rereads two icons of Canadian culture – *Anne of Green Gables* (Sullivan, 1985) and the Canadian songbird herself, singer Anne Murray – as sites for the inscription of queer desire. Where McLeod's mashup is an absolute departure from its viral *Brokeback* counterparts is in its insertion of her own messy commentary of desire, less than idealized expressions of female sexuality and dress on the original texts (Anne Murray's song "You Needed Me" and both Lucy Maud Montgomery's novel *Anne of Green Gables* and Kevin Sullivan's 1985 film adaptation) in her work, *That's right, Diana Barry. You needed me.*

Refusing to pass as straight, McLeod's work is remarkable and important for precisely that reason. T.L. Cowan states that McLeod's work "relishes, normalizes and hyperbolizes the unneutered realities and fantasies of all of us who don't [pass]." "[I]rreverent, unpredictable and twisted" (Cowan), McLeod not only interrogates the mainstream homogenized and (perhaps) desexualized 'good girl' Annes for expressions of their own sexuality, but imposes her own autobiography on their chaste, sanitized, and vanilla lives. Most importantly, McLeod constantly breaks the spectacular frame in her work by not allowing us to become immersed in the polished image of Anne (played by Megan Follows) and her friend Diana Barry (Schyler Grant), as McLeod remains in the corner of the framing, interrupting, jumping, and singing along with the media symbols that the two Anne(s) represent. McLeod calls her unique blend of performance and conceptual art 'interventions' (Hogan "Politics"), with addiction reference and the *double entendre* intended to be sure. In addition, this video is a *karaoke* video – we are invited to sing along to the lyrics that are printed as subtitles at the bottom of the screen. This is certainly not what McLaren had in mind when he sneered at the amateur sing-a-long nature of contemporary culture. McLeod draws us into the image, and invites us to assume the queer position, to become queer subjects. By recutting the *Anne of Green Gables* made-for-TV movie into an exploration of Anne Shirley and Diana Barry's lesbian love plot, McLeod highlights scenes where Anne gets her "bosom buddy" drunk, where they sleep in the same bed, give each other significant looks, chastely kiss, etc., where we can insert other meanings in a Twenty-first-century context. Anne as the iconic Canadian girl's role model is equally played for gags and problematized when McLeod says, "Now the one thing that Anne Shirley taught me how to do besides live my life as melodramatically as possible is to seduce girls with booze, booze, booze, booze, booze." McLeod thereby shakes the iconic image of the outspoken redhead to the core. She also does not simply reduce the *Anne* movie to a 1:1 site of fantasy, but complicates it by saying that Diana Barry ("you remember her, "Oh Anne, I'm so drunk [tongue, tongue, tongue])" was not an erotic object to McLeod herself, but that Diana Barry *reminds her* of her first girlfriend … whose name she cannot "say aloud for legal reasons … hers not mine." The mashup performs an inversion as McLeod reiterates that the film reminds her of her own 'not-gay' girlfriend who teased like Anne's friend. This ambivalent girlfriend, McLeod tells us:

> was all like "oh my god, I'm so drunk [tongue, tongue, tongue]. Oh my god, I've never been with a girl before [tongue, tongue, tongue]. I'm so not gay [tongue, tongue, tongue]. My boyfriend won't mind [tongue, tongue, tongue]." Yeah! 'Cause I'm really worried about your fucking boyfriend when I'm finger-banging you in the back of your station wagon after we've just had a night of Pictionary with my parents in their goddamned trailer!"

By recutting Anne and Diana with Anne Murray's "You Needed Me" as a soundtrack, she recontextualizes Anne Murray's 1980s frosted mullet and her own (and our own) adolescent memories of unspoken yearnings and hinted-at subplots in straight movies. By exploring sex versus romantic love in her video as well, McLeod goes far beyond queering the mainstream to queering the queer as she explores the inexplicable nature of sexual fantasy and the mysteriousness of what turns people on. T.L. Cowan argues that "cabaret has been a stage for poltically-motivated pastiche" since its inception in the early Twentieth century; in North American contexts, she argues, "cabaret might be understood as a mode of queer temporality that invites liminality rather than longevity, a kind of living that (in theory) thrives in the polyamorous (multiple) rather than the monogamous (single), and the low-paid rather than the lucrative." At the heart of Cowan's argument is the belief that "the outrageous, the gaudy, the raunchy" milieu of cabaret continues to exist outside of consumer culture parameters and so, therefore, can operate as a truly queer and countercultural space. McLeod uses this inclusive space as a site to construct a refusal to behave, a refusal to conform to the spit and polish of the mainstream stage or media machine – and part of that refusal, Cowan believes, is by resisting the drive for artists to clean up, grow up, professionalize, and produce a solo show. By releasing her video works out into the free economy of the Web however, McLeod does engage in a much-wider cultural conversation and greatly expands her audience and performance opportunities. Like McLeod's karaoke video, her performance is not polished or finished, but a site of navigation, a doorway, and a generator of our own queer participation. We too refuse the spectacular and become producers of desire within her shows as we sample her wares and try queerness on for size.

Remakes/adaptations/intertexts

In 1933 Director Leo McCarey filmed a scene with Groucho and Harpo Marx in *Duck Soup* in which Harpo plays Groucho's reflection in a mirror. (Legend has it that McCarey described what he wanted in the scene and that Groucho and Harpo then performed it perfectly in a single take.) Harpo, dressed as Groucho in identical clothes, is running to hide from Groucho when he crashes through a mirror. To stay 'hidden,' Harpo pretends to be Groucho's reflection. Groucho, suspecting subterfuge, is constantly trying to catch him out or trip him up, but Harpo always anticipates – until the end. At the end, Harpo is holding a top hat behind his back while Groucho has a summer straw hat. The tension in the scene is as we wait for Harpo to don the wrong hat – only he never does. Instead, he has a second hat behind him, and the gig is up only when Chico also steps into the scene dressed identically. This would have been doubly funny for the audience because the assumption is that the audience remembers the *original*. The original scene was in another Leo McCarey movie, which was filmed in 1924. *Sittin' Pretty* starred Charley Chase with an identical scene – except for the ending. At the end of that scene, Chase dons a different hat from his double. Fast forward a few years to May 9, 1955. Lucille Ball invites Harpo Marx to appear on *I Love Lucy*. Together, they perform the mirror scene live. The scene has a few changes, but it is clearly intended to be the same scene up until the end. At the end, Harpo and Lucy drop their hats, only Harpo's is on a string or a wire and bounces back to him while Lucy's falls to the floor. Which one of these scenes is the original? McCarey's remake of the first scene is far more polished than the original and transforms that material. Harpo's scene with Lucy is clearly a pastiche of the original as an act of homage. McCarey's original with Chase is clearly forgotten, and yet the most successful version is an intertext to the former.

Today remakes have remakes. While the Hollywood version churns out new versions of the same film over and over again, digital media artists and the free culture seek to speak with the original materials in ways that transform them so that they are entirely new. For example, Romanian artist Ciprian Muresan recreates Spanish Director Luis Buenel and artist Salvador Dali's 1929 silent, short surrealist masterpiece *Un Chien Andalou* shot by shot, but substitutes 3D animation for the original characters. Muresan's one-minute film with the same title (2007) re-creates the most famous scene where the young man slices the woman's eyeball with frame-by-frame perfection, only in his version the events play out between the ogres Shrek and Fiona. Presumably the artist's intent is to create a comic effect, but it is hard to know with such a bizarre translation of the horrifying original whether one is expected to laugh or cry.

Similarly Atom Egoyan remakes (or perhaps resituates) Federico Fellini's famous 'screen test' scene from the masterpiece *8½* (1963). In the original scene, a group of director Guido Anselmi's lovers see themselves and their histories play out on the screen as they watch the rushes from a burnt-out filmmaker whose inspiration has deserted him. The scene is a masterpiece study of the gaze. In Egoyan's remake entitled *8½ Screens*, he renders the scene into 9 (or 8½?) distinct points of view and combines them digitally, with each woman in a separate, fixed stream. He then projects the images backwards into a theater on to sheets hanging over the seating. The effect is electrifying and lends a whole new appreciation to Fellini's original. Also taken with Fellini's *8½*, Chinese artist Zhang Ding re-created Fellini's original film with a bicycle (instead a sports car) and much red drapery. *Great Era* (2007) is a stylized 14-minute single-channel video installation that has more to do with

Chinese opera than Fellini's original. While these works are wholly derived from the originals, they stand alone as well as being fascinating pieces in their own right.

Another kind of remake is the meme. One of the most successful viral YouTube videos is a 3:59 scene from Oliver Hirschbiegel's *Downfall* (*Der Untergang*, 2004). The scene is the one where Hitler (played by the legendary Bruno Ganz) learns that the war is lost and Berlin is about to fall. He begins by berating his generals for their incompetence and ends by contemplating suicide. The parodies explore Hitler ranting on about just any topic – from bands breaking up to the fact that he has become a meme. The smaller the slight, the funnier Hitler's rant becomes. In the wake of the takedown order in April 2010 from Constantin Films (discussed earlier) where most of these thousand spoofs were removed until they won their appeals (as permissible satirical material), more and more of the remakes were uploaded that were metatexts about copyright and fair use. These are four of the best. Part of the measure of their success is how well the new English subtitled dialog can be synced to the actors' lips. MastersofHumility's video "Hitler Is Fed up with All the Hitler Rants!" presents the scene in its original German with English subtitles as one of the Gestapo informs Hitler that news of his terror and destruction is being outshone by *Downfall* parodies on the Web. After ordering all social media users out of the room, Hitler then rants about the fact that the parody of him stopped being funny in 2008, responding to the idea that it's a "clever subversion of traditional media" by attacking the makers of the parodies and their effect on his online reputation. He says,

> I thought my legacy was secure. I slaughtered millions. Cut a bloody path of destruction across Europe. And for what? So I could be the latest juvenile Web fad? No better than YouTube Fred or that stupid f*cking hamster!? And they don't even edit the clip! (MastersofHumility, 2009).

In all the videos, the short exchange between the women waiting in the outer office becomes a commentary on the making of parodies. In this one, one woman says to the other, who is crying, "Don't worry. We'll post our vid anonymously." This parody speaks specifically to the perceived lack of originality in Hitler Rant remakes as well as the contagious nature of memes. This scene uses the actual legacy of Hitler as an allegory for, in this case, the massacre of creativity by copyright enforcement.

"Hitler, as '*Downfall*' producer, orders a DMCA takedown" was made by Brad Templeton (May 27, 2009). Templeton is Chairman of the EFF, the leading organization for the protection of digital rights. As Chairman, of course, Templeton was careful to ensure that his remix was entirely legal, but he reports at length on his blog just how complicated that was to do. Art imitates life in this video as Templeton casts Hitler as the Constantin Films producer of *Downfall* who initiated a DMCA takedown of all the *Downfall* parodies on April 20, 2010. In this clip, though his advisors warn Hitler that parodies are examples of fair use and that taking them down might "get [them] in legal hot water," Hitler orders every person who has ever violated a copyright to leave the room. While the subtitles inform us that the woman outside the office "cries over the death of free speech," Hitler ignores the Gestapo member who protests that encrypting everything would shut the "real innovators out of media." He continues his protest, "they always crack it and DRM just makes it painful to use." Hitler exclaims, "I don't care if it's broken. If it saves one DVD sale … Don't you grasp where our industry is? Our whole business model is selling copies. We can't let customers make copies!" Hitler then celebrates himself, exclaiming about how great Bruno Ganz (the actor) is as Hitler. "Digital copies are perfect. As good as the real thing. Better," Hitler says. Riffing on the argument that parodies increase sales, he says that they may as well all become Communists, "like Stallman." Richard Stallman, of course, started the Free Software Foundation, the GNU Project, and copyleft. He is a passionate campaigner against DRM. In the end, Hitler laments about the fact that the once-innovative film industry is now simply a "lawsuit factory." Ending with a decision to simply let YouTube police the parodies as a reflection of the deal with Viacom, this clip addresses the arbitrary and illegal nature of YouTube's copyright removal policies. Like MastersofHumility's "Hitler is fed up with all the Hitler Rants!," the video educates about intellectual property issues and parodies the actions of the studio using the very clip they are trying to block. The irony is that Constantin did a takedown of this video, and most of the others, on April 20, using YouTube's "Content ID" system. He disputed it and so it is back up again, for now…. Templeton's creative use of parody is a lesson in the major issues involved in digital rights management.

Dirtyfoxhead's "Hitler Pissed About Too Many Downfall Parodies" casts Hitler as the creator of the parodies themselves. He is upset because people no longer find his parodies funny and because too many of them have flooded YouTube with rants about banal content "from the collapse of the Mets to the traffic outside of Berlin." In this video, Hitler slams the unoriginality of much of the content on YouTube, including many of the *Downfall* parodies. The crying woman outside the office again represents the content creators on YouTube, but instead of being consoled on the grounds that her free speech is being infringed upon or that she can still create her content anonymously, she is consoled that "someone who hasn't seen these parodies yet will think it's funny." Hitler slumps into a depression:

"I always thought YouTube would serve as an outlet for creativity…. Instead, everyone posts FAIL videos or pointless video blogs. All shit. And it's all my fault. How is it that people finally got tired of Hitler? What did I do to deserve this scorn? Hitler out."

In "Hitler Rants about Downfall Parodies Being Blocked", hitlerrantsparodies deliberately targets Constantin Films, YouTube, and copyright enforcement in general by not presenting the video image of the clip at all. The screen is blacked out with the official message: "This video contains content from Constantin Films AG, who has blocked it on copyright grounds." The viewer is able to hear the dialog from the clip, but cannot see anything but the subtitles. Hitlerrantsparodies parodies the blocking of videos as an attempt to stop more parodies arriving. This video argues (as do many makers of parodies and other transformative works protected under fair use) that "our parodies never harmed Constantin. OUR PARODIES NEVER HARMED CONSTANTIN!!! These parodies promoted the film and helped improve awareness of the movies." While an unseen Gestapo advisor repeatedly asserts that the parodies are indeed a copyright infringement, Hitler speaks directly to the media coverage of the parodies and the legal battle that resulted in the takedown of the parodies on YouTube in April 2010 as well as to the subsequent increase in *Downfall* parody uploads following that event. Protesting that "a Downfall writer praised the parodies" and "the director said there [*sic*] hilarious," Hitler points out the irony in the fact that the parodies have been taken down while an entire copy of the film with correct English subtitles, presented in 17 installments, is left untouched.

This clip speaks perhaps the most clearly of the public and often reactive nature of Web culture:

> This attempt to wipe out the *Downfall* meme will only make people more determined to continue making them. They should have waited for it to die-off, eventually people would have ran [*sic*] out of ideas for new parodies. They have just made things worse. Now *Downfall* parody makers are actually disputing the blocking of there [*sic*] parodies. This might be the start of a world web war (Hitlerrantsparodies).

After the weeping woman – once again representing the repressed creative public – is consoled by her co-worker with the fact that "some are immune to being blocked," Hitler sighs because the blocking of his parodies is only going to ensure that his competitors succeed. The removal of the video feed for the entire clip underlines the idea that blockage does not stop parodies. Parodies are free publicity for the film industry in general, and this version parodies that fact by using Constantin's own notice behind the subtitles. As Bourriaud (2002a, 19) observes:

> the contemporary work of art does not position itself as the termination point of the "creative process" (a "finished product" to be contemplated) but as a site of navigation, a portal, a generator of activities. We tinker with production, we surf on a network of signs, we insert our forms on existing lines.

These videos are not the end of anything. They are contact points where the baton gets passed from one artist to another.

Scott Kildall's *Window* (2007) is another example of this kind of recursive looping remake of an original work. Unlike his other performance art pieces, Kildall remakes a film here, a film that is itself a remake. Using Pierre Huyghe's *Remake* (1994–5) as a model for re-creation, Kildall's performance and photograph restages Huyghe's video restaging Alfred Hitchcock's *Rear Window* (1953). Only that is not precisely what Huyghe undertook. As Michael Rush puts it, Huyghe is "Not a typical 'appropriator'" (Rush, 2007, 170). Undoing or deconstructing the traditional Hollywood remake, Huyghe starts from scratch with his own costumes, sets, and actors

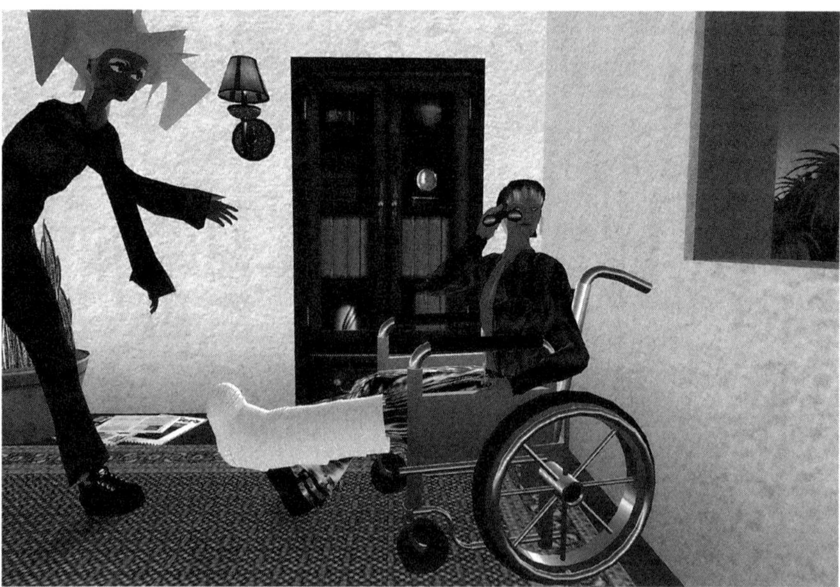

Figure 6 Scott Kildall, *Window*, and Pierre Huyghe, *Remake*

to expose the structure of the original film by de-emphasizing the mystery of the narrative. His actors have difficulty remembering their lines, they behave in a way that does nothing to retain the tension of the original. Their very distance from the storyline allows Huyghe to expose Hitchcock's methods of cinema (Rush, 2007, 170).

Huyghe's actors, sets, and costumes make no attempt to copy the Hitchcock film, only to set a comparable scene. In the same way, Kildall's *Window* relies only on the props of wheelchair, binoculars, and window. Kildall himself stands in for the Grace Kelly character with his large, bizarre avatar with geometrical pink hair, purple skin, and dayglo clothing and another character is in a wheelchair. In fact, if the C-Print were not part of a series restaging famous performances we might miss the allusion altogether.

The *pièce de résistance* in digital performance is Talan Memmott and Eric Snodgrass' piece *Intermission* (2011). They call it a "performative re*dada*ction of the poetics of cinema" (authors' description). A grand remix of the unconscious of cinema, René Clair's *Entr'acte* blends and bumps up against Picabia and Satie; it is "performative cinema entr'activism – a form of participatory filmic event in which cinema is mobilized as the physical borders of frames leak back into a precarious hyper-reality that emphasizes the metanoic qualities of being in-between acts of viewing" (authors' description). Those so-called 'metanoic' qualities are the veil of participation that draws in the audience, redirects them, and unites them in a common vision. In the metanoic realm, each individual knows they are a part of a group, but are empowered to realize their own vision even as they empower those around them and the society at large. By sucking the passivity out of the screen, *entr'activism* refuses the passive and the active, and instead cycles within a perpetual state of flux. A *détournement extraordinaire*, *Intermission* is always a direction, never a goal. As Dziga Vertov's *Man with a Movie Camera* cycles within and behind the whole experience, the cinema is interrogated, buried, and resurrected as a new form within the ecosystem of screen culture.

Streamed data/content or visualization

The closest thing we now have to a live experience in our mediated age is a streamed experience, according to Christine Paul (2008, 18), and since performance- or time-based work exists only in its documentation streamed data, captured content or visualization are become primary tools for art and for information. Capture can be as simple as the viral YouTube video *Noah Kalina Everyday,* in which this photographer made a time-lapse film out of the single picture he took of himself every day for six years. Or it can be as complex as Manu Luksch's *Faceless* (2007), a 55-minute movie (starring Tilda Swinton) filmed over a period of months using CCTV surveillance equipment. *Faceless* explores an era of calendar reform in which the past and the future have been eliminated from people's lives. We could call Luksch's film *Postcinematic Surveillance.* Another kind of postcinema is *Self-Surveillance* (or inverse surveillance or sousveillance). After a traumatic six months of interrogation and investigation by federal authorities followed a case of mistaken identity for Bangladeshi-American artist Hasan Elahi, he pre-empted further investigation and identity confusion by documenting all aspects of his life. Wired with a GPS unit and photographing himself, his food, and his surroundings on a regard basis, the FBI (or anyone else) can track him live in real time at any time. The pictures at his site number in the thousands as he uploads more than 100 photos a day. *Tracking Transience: The Orwell Project* (2002 on) is the result of a cell phone hack that links his mobile to the GPS network and records his location and masses of personal data in real time online. *Paranoid Scopophilia* is another datafeed postcinema genre. Digital border surveillance cameras do not see images, but instead track data. American Jordan Crandall's *Heatseeking* reroutes military cameras as a site for staging s/m panoptic violence and erotic encounters outside visible zones. Bodies are weaponized through the militarized machine-image complex, and eros and paranoia become intertwined. Over the course of seven films, blending a combination of film and video surveillance footage, stealth technology, and miniature infraread capture systems, Crandall explores the high-tech and aggressive policing of the San Diego–Tijuana border. The gaze goes both ways, and Crandall examines the nature of voyeurism and control. He says:

> The 'border' is not only a territorial marker but a provisional divider, helping to contour self and body, and its policing mechanisms have subjective dimensions. Tracking, targeting, and identifying formats begin to seep into the way we see, behave, and desire. They enter into the very structure of perception. The camera marks the place of battle (Crandall, 2000).

Surveillance art can also blend live performance with surveillance and the news genre as in Rod Dickinson's collaboration with Steve Rushton, *Who, What, Where, When, Why and How* (2009). Similar to the ambiguity that is found in Mez's work, Dickinson and Rushton leave us feeling unsettled as they simulate a press conference with "The Politician" and "The General." They read found texts from political speeches, designed to justify state-sanctioned acts of aggression, delivered by heads of state since the Cold War. They all sound eerily alike – regardless of whether they were uttered by benevolent ruler or bloodthirsty dictator. A powerful indictment against politics and war, Dickinson and Rushton set out to explore "the feedback circuitry of television" and how it "shapes social and political reality" (Dickinson and Rushton, 2009).

Archiving as an aesthetic form

In the new media, archiving and searching start to become a form of creative practice. They are not part of the composition, but the process by which an end-product is generated. An encyclopedic search was enacted by Alonzo Mosley, a Florida librarian, in *100 Movies, 100 Quotes, 100 Numbers*. It mashes remixed movie clips of iconic film stars in their greatest roles speaking numbers from 100 to 1 in 9 minutes, 28 seconds. It has such a hypnotic effect for the viewer that *Guardian* film critic Peter Bradshaw asked if it might be "the greatest YouTube clip of all time":

> a mind-blowing effort of archive research, somehow trivial and monumental at the same time. The effect is brilliant, hilarious, even weirdly moving. The final ten clips have a tension and a crescendo of their own, as you try to guess what they're going to be. I should have sussed the final one, but I didn't. It functions as a mini-masterpiece on its own terms and also as the basis of a fantastic pub quiz round, in which contestants have to identify each clip (Bradshaw, 2007).

One highbrow gallery equivalent of this is the London-based duo Thomson & Craighead's installation *The Time Machine in Alphabetical Order*. It resequences the whole of the 1960 George Pal film that starred Rod Taylor alphabetically by each word of dialog. Another gallery example of this kind of art is the work of another British artist, Alistair McClymont. His work is a nine-monitor installation called "Art for Art's Sake" that plays all of the many versions of the Metro-Goldwyn Mayer logos – a roaring lion's head with the motto "Ars Gratia Artis" – that were used over the years at random.[6]

What computers have turned out to be best at is searching. What they are worst at is archiving. Obsolescence is built in and inevitable. In *The Secret War Between Downloading and Uploading*, Peter Lunenfeld talks about the aesthetics of the unfinished in digital spaces (136) as process comes to the fore. According to Paul Valéry, "All work is beta." For Valéry, a poem is never finished, only abandoned. Former products are increasingly becoming mere content – or, more exactly, data – in this new process of assembly, play, and performance. The Australian remix duo Soda_Jerk (comprising Dan and Dominique Angeloro) says that

> the randomness of sampling means that you can't help but develop a kind of obsessive gambling mentality. It's addictive. Sometimes you might go through ten videos in a night without finding anything of use, and another time you might land a few gold samples in a period of minutes. If we faced up to the insane amounts of wasted time involved with sample-hunting, we'd never be able to go through with it. So instead, we endorse the delusion that the jackpot is always just about to drop (quoted in Harley, 2009).

The process is the experience, not just of creating digital media, but of always already remaking digital media. The protocol of the search becomes not a part of the work, but the process of the work itself. Conceptual art used computer protocol, still at its beginnings (the products in question would not truly make their public appearance until the following decade.)

Hacks

I have talked about a lot of hacks in these pages, but I wanted to call attention to a few more because the hack is such an important model for authorship in new media. One of the most famous art pranks was carried out by cyber-feminist Cornelia Sollfrank. In 1997, a call for works went out for a show called "Extensions" at the Gallery of Contemporary Art at the Hamburg Art Museum. Sollfrank hacked the application with an overall work she called "Female Extension" (1997) and applied under hundreds of female alter-egos located in seven different countries:

> Sollfrank drily countered 'Extension' into the Net with a 'Female Extension': she created 288 international female artists, complete with postal and e-mail addresses. Sollfrank generated individual Net art projects for 127 of these artists, using a computer program that collected random HTML material from the WWW and recombined it automatically. ("Sollfrank", *Media Art Net*)

According to Shanken, the organizers were thrilled with the popularity of the show and the diversity of the applications – of more than 280 applications, two out of every three were female. Nevertheless, the three top prizes went to men, and the jury did not understand the magnitude of Sollfrank's contribution (or did not read her press release which explained the hack), which she called "'the apparently meaningless flood of data' produced by her 'automatically generated net art'" (Shanken, 2009, 36).

Two other hacks are also worth mentioning here that both interacted with Amazon, the online bookstore. *Amazon Noir* was a major pirate event that was both a technological and conceptual performance that ended in November 2006. As the name suggests, it was a crime story in which the hacktivist team Ubermorgen.com (Paolo Cirio and Alessandro Ludovico) stole books from Amazon. They did so by repurposing Amazon's search engine, which will allow you to search for 50 words out of any book at a time. From those 50 words *Amazon Noir* would begin to reconstruct each book it selected from front and back simultaneously, one paragraph at a time. An activist response to the criminalization of downloading, they called their backdoor bot a "robot-perversion-technology" and it was connected to:

> four servers around the globe, everyone with a specific function: one in USA for a faster sucking of books, one in Russia for injecting books in p2p networks and two in Europe for schedule (*sic*) the action with intelligent robots. The main goal was to steal all 150,000 books of the Amazon's "Search Inside the Book" feature, and then use the same technology … [to] … steal books from the Google Print Service (Thalmair, 2008).

Modeling themselves after the digital resistance of the Critical Art Ensemble, they sought to collapse the gap between art and life to make books available for free download. The books, however, were not camera-ready copy. They were reconstructed 50 words at a time and so they were missing layout features and images – often making them hard to read (Thalmair, 2008). Ubermorgen see rematerializing art as a trend, and they seek to rematerialize digital paradoxes and destabilizing potential markets in a 'conceptual economy'" (quoted in Thalmair, 2008). The process of rematerializing was to apply print on demand (POD) technology to redistribute these works in poor countries and to question copyright restrictions (Thalmair, 2008). When Amazon eventually realized what Ubermorgen were doing, they purchased the program outright from the duo in a lucrative deal.

The second Amazon hack was called *Pirates of the Amazon*. *Pirates of the Amazon* was "an artistic parody" run by the Piet Zwart Media Design MA Program in Rotterdam, The Netherlands. It was designed to plumb information access, information design aesthetics, and issues of piracy in contemporary culture. Where *Amazon Noir* stayed up and running for months undetected,

Pirates of the Amazon was detected within a day and ordered down by Amazon (December 9, 2008). What it did was search the piratebay.org bit torrent for titles of books available on pirate sites, and then it transformed the center of the Amazon page into a "Free Download Button." It did not actually permit you to download material for free so much as tell you that it was available out there. It was not so much illegal as misdirecting. It took you, eventually, to illegal sites. It was a great hack and reversal on Amazon's own advertising policy: which it had been "doing the opposite for years now – placing ads on torrent-sites and the like, where you can buy the same item from Amazon you are about to download for free. Therefore 'parody' is indeed the right term to use for this plugin" (*Pirates of the Amazon*, "documentation").

Google Empire: Smart art, intelligent agents

For the past few years, visualization has been heralded as the next 'big thing,' and it is a thing that is already upon us. Information architecture – the rendering of information visually – is a way of modeling the world around us through the organization of time and space. Increasingly, as our world becomes more and more informational, the old hierarchical or vertical methods of organizing information (such as tables of contents, lists, and so on) no longer provide enough complexity to cope with the masses of data we now sift through on a moment-by-moment basis. Visual indexing not only helps make sense of this glut by organizing information associationally and spatially, but has also rapidly evolved into a medium of expression for art. The new engines that perform complex acts of information modeling (like Pivot) have the advantage of being networked, adaptive, and interactive, allowing for both rich results and aesthetic expression from forays in information foraging.

Literary critic Jerome McGann has observed that experimenting – what we might think of as the conceptual modeling of ideas – is about "imagining what we don't know" (quoted in McCarty, 2003, 52). The act of finding ways of visually representing information, in other words, reveals new kinds of knowing, just as art has always done. The conceptual modeling of ideas is something that spans all disciplines and cultures, but the maps that are generated there are rooted in their own time and space, and echo their originary cultural frameworks. While the dawn of civilization coincided with the invention of the first tool, the quest for intelligent tools has been the driving force behind innovations as diverse as social movements, education, and industrial design – from self-aware and self-directed tools like slaves to battery-operated educational toys that automate our children's learning to the programmable and interactive tools produced by the robotics industry. The ability to learn or adapt cognitively, physically, and emotionally are signs of intelligence; now that our electronic tools can possess or seem to possess these capacities, these attributes are being harnessed for artistic ends. In our time as datasets as complex as the human genome are visualized, one of the most pervasive searches is for tools that make the visual indexing or conceptual modeling of information possible. However, where the ancients drew these models in aesthetically interesting ways, our spatial models are able to realize these shapes in such astounding complexity that they not only aim for aesthetic effects but exhibit 'tool intelligence' as well. This issue was at the crux of computing pioneer Alan Turing's investigations; his driving motivation was the question of whether "computers 'could think'" and, if so, how that might be compared to human logic (Greene, 2004, 16). When these information models and engines are used for something more than utilitarian ends, they move into the realm of smart art and intelligent agents – that is, generative art or art that generates its own original works and aesthetic effects. This raises the question: What exactly is intelligence? And, as a corollary, how do we measure its presence when it pertains to our tools and our art? Benjamin Bloom's taxonomy of the three domains of learning – three different kinds of intelligence – might serve as the most useful model here.

According to Bloom, the cognitive domain is that of mental abilities or the realm of fact; what is considered knowledge. The psychomotor category comprises manual or physical skills: physical intelligence. The third area, the affective, is the emotional sphere – what Bloom called attitude. An exploration of a few recently developed tools that gather, filter, and model information in ways that seem to demonstrate some or all of these kinds of intelligence will highlight exactly what intelligent tools and smart art might look like. It is my contention, however, that we do not recognize these objects as intelligent, and that they do not become compelling, until they start to demonstrate what we deem to be an emotional response.

Google is the most popular and the most powerful search engine on the World Wide Web. While it does not undertake the *visual* modeling of information, it is an engine that seems to exhibit intelligence of a sort through its selection criteria. More than this, since it is one of our most familiar ways of engaging with both information and the Web, it is useful for demonstrating ways in which information can be rendered. This is important because technology fundamentally changes any tasks we undertake and modeling itself is difficult to conceptualize. In 1998, two students at Stanford University,

Google developers Larry Page and Sergey Brin, developed an algorithm called PageRank. It revolutionized all existing notions of search engine capabilities by adding a statistical component that could rank the pages on the basis of preset criteria. Those criteria are threefold. They are popularity, page attributes and content analysis (Brandt, 2002, n.p.). Content analysis "generally takes the form of on-the-fly clustering" of relevant content "into two or more categories, which allows the searcher to 'drill down' into the data in a more specific manner" (Brandt, 2002). While we get only a hierarchical text-based printout of Google's findings, what quickly becomes visible is the very anti-democratic nature of page ranking. For example, it favors pre-existing pages over new ones, privileges large corporations with a heavy presence, and often does not count small sites at all (Brandt, 2002). Telling evidence of this is the fact that Google claims to be comprehensive, but no one search engine even grazes the surface of the Internet's estimated (in 2002) at more than 550 billion documents ("Search Engines," 2002). Google, the most thorough, by 2002 only sampled some 2.5 billion pages ("Search Engines," 2002): approximately .5 percent. Google has not published the size of its index for years now, but in 2010 search engine expert Borslav Agapiev stated that of the then estimated 120 billion URLs in existence Google had indexed at most 12 percent of the open Web. More telling than that though are the increasing numbers of unvisualizable, dark areas on the Web that have been spawned by privatization and proprietary sites like Facebook and craigslist.com. Referred to via evocative metaphors like 'moated warehouses' and 'walled gardens,' the Dark Web or Deep Web consists of tens of billions of uncrawlably formatted, unindexable pages concealed behind 2.0 software (like comments on blogs), subscription walls (including large-scale works like *The Orlando Project* and most academic journals), and form-driven access to networked database directories (like the Library of Congress).

Information surrounds us and is now stored spatially rather than being written. Like time, it is largely invisible to us, since it has acquired a kind of transparency. Modeling can help make us more aware of its attributes. We now see that information is not at all the linear or analog form that was conceptualized by Claude Shannon in the early days of the telephone, but is instead a noisy matrix or network incorporating all perspectives. It is how we communicate with each other in an age of broad bandwidth technological mediation, and, on the World Wide Web, how we move through information space as subjects surfing with a mouse is what defines the environment's shape. In earlier times, knowledge was conceptualized structurally such as in the Jewish Cabala; in our time, knowledge is often drawn to be beautiful but useless. (See Johan Wastring's *Visual Information* (http://www.visual-information.org/), Manuel Lima's *Visual Complexity,* also online (http://www.

visualcomplexity.com/vc/), or Martin Dodge and Rob Kitchin's exploration of the geographic metaphors of the Web in *An Atlas of Cyberspace* for a wealth of examples. (The latter is accessible online at http://www.cybergeography. org/atlas/geographic.html.) Once information is aestheticized it gives us an overall perspective or single image, but generally little hard data – unless it is a very large-scale work like Microsoft's Pivot[1] or Barrett Lyon's *Opte Project*.[2] Until very recently though, such technology could not even have been imagined. In 1998, the British trio I/O/D 4, composed of critic Matthew Fuller, programmer Colin Green and artist Simon Pope, played with these concepts when they transformed a browser into software art. *Web Stalker* was very influential when it was first released in the 1990s because it visually mapped data in conceptual categories (Greene, 2004, 86). Deliberately playing with the cognitive mind behind conceptual modeling, they were crusaders against commercial browser software that ignored user wants and needs. The trio's moto is "Software is mind control – get some" (I/O/D 4).

Clearly the new electronic means of information rendering are shouldering a large portion of the paradigm of cultural representation and can help us gain perspective on the crushing burden of information that is piling up around us. In the aestheticized architecture of the Information Age time and space begin to merge, becoming something visual that we move through. This may be self-evident now but when Matt Kirschenbaum wrote in 1998, in "A White Paper on Information," that information has come to acquire meaning it was a statement that could still surprise. Kirschenbaum went on to say that information has become increasingly aestheticized and assumed a recognizable form in recent years. Data is now modeled and rendered visible in identifiable structures and shapes: "at precisely the moment data becomes invested with visual form as information," he says, "so too does it assume a cloak of representational artifice, thus taking its place in the multifaceted media array that has defined the popular contexts of the Information Age" (Kirschenbaum, 1998, I). In the new media, information modeling is the measure of our surfing in virtual space and the accounting of our journey. Time in information space and in electronic texts is loopy, following vectors and flows, and folding-in places that are revisited again and again. There are no linear lines here, since 'forward' and 'backward' have no meaning when the only constant point of reference, the only situated perspective, is oneself. This disorientation is endemic to surfing the Web; hyperlinks flip a browser to and fro in virtual space and time.

Oral cultures are masters at modeling information in visual ways. Perhaps the greatest achievement of conceptual mapping was the ancient Art of Memory, a visual method of memory storage. Practitioners were so skilled that they could be what we would consider the first computers. The Art was originally used for oratory, and memorization was kinetically encoded in the body by walking through the chosen space. The Art continued to be practiced in the early days of literacy and was fine-tuned to include verbatim recitals by master practitioners of, for example, the works of Aristotle or the Bible. In the Middle Ages, the Art came to be adopted by followers of the occult arts, alchemists, hermetic philosophers, and Jewish rabbis governing Cabalism who developed their own visual models. It was among these uses that Giordano Bruno and others embarked on a quest for the "universal memory machine" (Yates, 1992, 206), and the search for the divine plan of the universe became a hunt for devising a system for modeling all human knowledge. Bruno, for one, sought to harness the omnipotent magic of the heavenly bodies in highly systematized memory machines. These cosmological machines had little of the mechanical about them apart from clockworks and the concept of turning concentric circles. Instead they were primarily informational. They consisted of attempts to organize the secret teachings of divine knowledges on conjuring wheels that would create a model to capture the holy powers of the universe, and subsequently endow the speaker of 'magical' words with mystical powers. Frances Yates in *The Art of Memory* argues that these alchemically driven endeavors to map the magical powers of the universe in the Renaissance ultimately led to the concept that the universe was a mathematically driven piece of machinery: a giant clockworks. This concept would ultimately be turned back on mortal man, to study first the mechanics of the human body and then the brain as an information processor, a prototypical computer, or intelligent tool.

As literacy spread during the Renaissance and Restoration and printing made texts accessible, the need for the instant recall of massive quantities of fact started to fade. In its place, new informational models began to be devised in print. Encyclopedias, dictionaries, and grammar texts were born, and the word began to supersede the image as the preferred system of memorization. As a result hierarchy began to be favored over association and "spatial visualization start[ed] to take place on the page" (Yates, 1992, 230) typographically rather than in the mind. Bruno wished to devise a method for categorizing all of human knowledge, but it was not until the 17th century with its preoccupation with a universal language that a viable answer presented itself – in the truly international modeling language of mathematics (Yates, 1992, 364). Philosopher and mathematician Gottfried Leibniz's (1646–1716) goal was not so different from that of his predecessors: along with his new language, a notational system for infinitesimal calculus, was an encyclopedia or a compendium uniting all known facts in the arts and sciences (Yates, 1992, 368). Leibniz referred to his calculus as "a true Cabala" and "a universal key" (Yates, 1992, 371). More importantly, Leibniz freed logic, and specifically symbolic logic, from the restraints of natural language, and introduced a notational, numerical, information-based model that could talk across systems and cultures. It is a system that we now know better as binary code, the language of computers.

In conceptual models of systems of knowledge (what we might consider models of complexity), the shape of the infinite is depicted as concentric circles. This shape is most familiar in Dante Alighieri's (1265–1321) representation of the three celestial planes of Paradise, Purgatory, and Hell in his *Divine Comedy*, but it also rematerializes in the most important mathematical model of infinity visualized in the 19th century: Georg Cantor's infinite levels of infinities nested within infinities, the transfinite numbers or alephs (Aczel, 2001, 69, 140–8). The aleph is a point in space that contains all other points (including time). Cantor's ideas are probably most familiar to literary scholars through the blind librarian Jorge Luis Borges's exploration of the infinite in a short story "The Aleph." After peering into the aleph, Borges' fictional self struggles to describe the information overload of his glimpse of the entirety of the visible universe. This is the inconceivable infinity that, he observes, mystics evoke symbolically to describe the nature of the Godhead – and that so often drove all those who sought to define infinity, from rabbis to mathematicians, mad.

A Cabala is one such knowledge system; it is a networked, topological model and a means of navigation through a divine cosmology. Through magic and astral travel, the Cabal is a projected journey through three worlds and ten dimensions. Part of the significance of the Sephiroth is numerical, with the word literally meaning "countings" (Aczel, 2001, 32). Its ten elements are arranged like spokes on a wheel with the sixth element, the aesthetic dimension 'beauty' at the center as the connector. The drive to understand the model or the infinite connectedness of all things was seen as a holy quest, and, like the vanishing point in a Renaissance painting, the Sephiroth were a model of the invisible and unattainable existing beyond the horizon of God's immortal plan (Aczel, 2001, 43).

Michel Foucault is known for his passion for mapping the foundations of particular types of knowledge. In *The Order of Things*, Foucault seeks to construct a visual model for the contours and coordinates of the episteme of knowledge, those foundations of classification and organizational structure that are integral to systems of thought. Specifically he looks to identify the importance of the visual science of 'resemblance' and the four key 'similitudes' in the Western world as they mattered and ruled from the late medieval period until the dawn of the age of Reason. The web of resemblance that he charts – through its components *convenientia, aemulatio, analogy*, and *sympathy* – is a matrix of associational logic and connectivity similar to the

World Wide Web, but that is linked not by hyperlinks, but by relationships in language. *Convenientia* is fixed but linked adjacent space; it designates a relationship between things or ideas (Foucault, 1994, 18). *Aemulatio* is mimicry; it is emulation freed from location and connection as a kind of simulation: "it is the means whereby things scattered through the universe can answer one another" (Foucault, 1994, 19). *Analogy* is the superimposition of *convenientia* and *aemulatio*; it connects resemblances across space and time, and simultaneously links man to the rest of the universe (Foucault, 1994, 21–2). Finally, *sympathy* is a dynamic principle of transformation, defining mobility and interpenetration; sympathy transforms differences into similarities, rendering connection as a gesture. These four methods of classification demonstrate how "the world must fold in upon itself, duplicate itself, reflect itself, or form a chain with itself so that things can resemble one another" (Foucault, 1994, 25–6). This is not the end of the story though, since these methods must be rendered into the system of signatures.

Signatures alter the relation of the visible to the invisible – perhaps rather like the relation of the virtual to the real – by defining the relationships between things. Similitudes must be 'read' (or their secrets divined) and hence signatures set out to map these elaborate interconnections in language. Similitudes are not to be confused with comparative biology or other scientific bases for comparison. For instance, a plant with a resemblance to a body part would be used to treat that part's diseases, as the walnut, for instance, with its hard shell and wrinkled interior, was used to prevent internal problems with the human brain which it resembled (Foucault, 1994, 27). The divination of resemblances was a visual art and spiritual practice. The sole scientific method was in the recording of observations in print. Writing, therefore, comes to the fore in the Renaissance world picture – contemporaneous with the rise of the printing press – as the most important medium of information storage.

According to Foucault, every resemblance has a signature, what we might think of as an icon or trademark or breed trait in contemporary terms, but this signature is simply relational and therefore an alternate form of resemblance of the same thing. It should be immediately clear that this system has information overload as a precondition. Ultimately, if one looks hard enough, one will begin to see resemblances everywhere. Everything in the microcosm will ultimately resemble the infinite macrocosm itself. Resemblance is a model for similitude, an enormous interwoven network of links. The drive to record and remember the world in encyclopedic detail is born of a need to 'see' the information. The worldview as a category of thought interweaves duplicated resemblances with a macroscopic justification on an ever-larger scale. This rendered everything classifiable – all documented resemblances – within a form that duplicated the shape of the cosmos (Foucault, 1994, 31).

In the past ten years, Google implemented a similar system on its innovation lab, labs.google.com, described by the company as "Google's technology playground." This is a showcase for ideas not yet ready for commercial release. One of these search features is called Google Sets. It creates groupings of similar items: "For instance, type in a query for Donner and Blitzen, and you'll get links relating to all of Santa's reindeer; type in Harvard University and Stanford University, and you'll get links to other top-ranked schools" ("Search Engines," 2002). Craig Silverstein, Google's technology director, says the sets feature could also be used by job seekers wanting to find similarly named companies ("Search Engines," 2002) or by anyone who is searching for information that has associational connections. Clustering, tag cloud, and classification metasearch engines like Quintura and Yippy (formerly Clusty) also help visualize the relationships between things.

The proto-Renaissance thinker, like the Google searcher, was caught in "an endless spiral" (Foucault, 1994, 32), drawing knowledge and sources of all kinds into the model. Everything needed to be included because it was the resemblances that mattered – rather like our contemporary privileging of patterns and connections. Robert Fludd's Great Chain of Being (*circa* 1617 to 1621), for instance, was designed as a representation of the cosmos. These compulsive cataloguers sought to re-create the divine order and original plan in infinite endless compendiums just as we seek to include all knowledge in webs of previously unimagined complexity. Writing to them was a part of nature – given by God – and therefore never arbitrary. It provided a direct link to the secret model of ancient, magical, and holy languages. Alchemists and hermeticists sought to find orientation in the direction languages were written from and in, as a literal map of the cosmos. The tableaus and compendiums of their knowledge were designed to re-create the divine plan of the heavens in the perfect form of the circle and the branching shape of the tree. Rhizomatic and networked in nature, this was the topological theater of information that the Renaissance alchemist modeled using interactive spaces rather than merely recording and remembering the facts in print. These systems "spatialized acquired knowledge" as both information trees and magic mechanical wheels (Foucault, 1994, 38).

The need to cognitively, physically, and emotionally model information – and specifically to do so in visual ways – is as old as civilization itself. While the methods, rationale, and reasoning have changed dramatically, the impulse remains constant: to devise a system for graphically and spatially organizing large quantities of information to make it easily navigable. Visual representations can be understood at a glance, while numbers, pie charts, and graphs must be meticulously analyzed and assessed since they focus only on the cognitive. Hierarchical or vertical structures are still commonly used in electronic spaces, but as a form of organization that had its origins in print this is considerably off the mark in a time when *Web Stalker*'s kind of visual indexing is what people want. The more logical way of structuring in the electronic realm is to use a nonlinear, digital-native form: associational or horizontal linking. This latter form also includes elaborate structures such as map-based navigation. Associational linking is more difficult to alter or add to as it evolves than a hierarchical site, which is one reason why it is less popular with designers, but it is also capable of greater subtleties and nuanced effects. In doing so, it also raises the potential for bringing the full spectrum of Bloom's KSA scale into play. In essence the horizontally linked structure is an information space that privileges aesthetic effects, scalability, spatial navigation, and often demonstrates tool intelligence as well. In stark contrast to Google Sets, for instance, there are also Google Goggles for the Android phone. Take a picture of anything – a painting, a bottle of wine, an intersection – and the app uses place matching, optical character recognition, and object recognition, and it returns relevant, related links. Is this intelligence? The next step up the ladder at Google Labs is Gesture Search. Draw a letter (for English words only) on the touch screen of your Droid and that will call up contacts or apps or bookmarks or other files which contain that letter or combination of letters. This is starting to get eerily smart.

The ever-increasing quantities of data in our daily lives call for ever-more complex maps and ways of mapping information multidimensionally – that is, both horizontally and vertically in time – across multiple trajectories. Older, print-based styles of linear organization just do not yield enough complex data to satisfy our hunger for information any more. One multidimensional carto-graphic model is marketed as "Business Intelligence Software"; Antarctica's *Visual Net* provided big picture perspectives. (This was developed by Tim Bray of Vancouver – also one of the developers of XML). The website claims: "With Antarctica's *Visual Net*, *information consumers* are able to 'answer questions they didn't even know they had' and identify problems before they become critical issues" (formerly cited at http://antarcti.ca/products/visual_net/; the content has now changed). Creating a custom, map-based interface of a customer's data, users can navigate this "visual landscape" where solutions can be 'seen.' The visual nature also allows critical elements to be grasped at a glance. Since the software presents "multiple dimensions of a problem … consumers gain a complete understanding of an issue, unlike" more conven-tional tools that depict these issues in only two or three dimensions (formerly cited at http://antarcti.ca/products.html). Like a topographic map, *Visual Net* shows the geographic significance of particular topics. Each category can then be zoomed in on for navigation down through further subcategories. This is a useful tool to be sure, but hardly what we would call intelligent.

Anemone is smarter. Where *Visual Net* simply constructs zoomable maps of content, *Anemone* builds maps of website usage. Organic in nature, it can grow and atrophy as it charts its browser's travels. Developed by Ben Fry at MIT's MediaLab as an example of "organic information design" derived from large, dynamic datasets, *Anemone* is an autonomous agent that is similar to an organism (Fry, 2010). Visiting a page produces a branch that begins to grow and that thickens with each subsequent visit. Its smarts lie in this measure of usage to determine which branches will grow and which will atrophy and wither away. Thematically connected areas of study also grow closer to each other based on predetermined growth and movement rules (Fry, 2010, n.p.; Dodge, 2004, n.p.). This is organic information design that allows us to see site traffic at a glance, like a snapshot, in a way that we could not process with graphs or charts. For this reason it does not depict the whole site, only visited pages. Fry says that a challenge for him in designing it was to resist the temptation to design a simplex map or a purely aesthetic object. Instead it is a software program that eats data and, as a digestive act, reacts to it. This reactive nature makes it appear intelligent. Fry claims that he did not set out to play Victor Frankenstein (although his later work is concerned with visualizing the human genome), but instead uses attributes that living organisms exhibit (e.g., growth, atrophy, and metabolism) to give it the semblance of life: "watching how the organism responds tells me what's in the data.... Even the simplest organisms know how to respond to complex stimuli, so making a visualization based on its traits has potential" (quoted in Dodge, 2004). What is key here is not just having a tool that will map data, but the extent to which this tool may be used to generate smart, personalized maps of the data we need or want to navigate.

Similarly, *Grokker2* was (its developer Groxis ceased operations in 2009) productivity software designed to render customized maps of data. What *Grokker* had that *Anemone* lacks was the ability to filter, edit, save, and share the results of mined data from a user's own computer or network or database, or from the Web. When used for extrapolating info from the Web, it harnessed seven of the major search engines – AltaVista, MSN, WiseNut, Fast, Yahoo, Teoma, and Google – to generate subject maps of up to 2000 items. This is how it worked: if your topic was a city, it generated political, historical, and geographic categories, as well as museums, universities, hotels, tours, major monuments, and so on. If you entered a term like Cubism, it would break the topic down into subcategories that include Picasso, painting, other artists, and web resources. You could plug in filters on the fly, narrowing the search to words, to a number of ranked sites of your choosing or by Internet domain. Then you could edit and save these maps for later reference. For instance, this smart tool could produce snapshots of your notes toward your dissertation or latest book as they grow over time.

KartOO (2001 to 2010) went still further. It was a meta-search engine that visually modeled product-oriented information about information retrieval. Launched in 2002 by cousins Laurent and Nicolas Baleydier in France, it produced its results as interlinked document icons sized according to their relevance. *KartOO* was a Web-based flash interface. A user entered a search term, clicked okay, and "*KartOO* launches the query to a set of search engines, gathers the results, compiles them and represents them in a series of interactive maps" (Dûrsteler, 2002, n.p.). What makes *KartOO* especially interesting though is that the engine draws semantic links between results (Dûrsteler, 2002, n.p.) shown as snaky lines connecting documents. Semantics are increasingly being taken into account on the Web as the quest for standards appropriate to making semantic linkages continues. But semantics use ambiguous language or notoriously fuzzy logic that is not normally deemed what Tim Berners-Lee calls "machine-understandable language" (Berners-Lee, 1998). While the supporters of the Semantic Web are known for making overly large claims about its potential, the capability to search and map across a cacophony of metadata standards, namespaces, schema, languages, and ordering systems is urgently needed. Well-designed information architecture will help smooth this transition, drawing content, context, and users into the equation, but in the interim *KartOO*'s kind of visual indexing helped point the way. The semantic dimension of *KartOO* functioned almost like a narrative, and we could follow these information trails to narrow or expand our search. Where traditional search engines like Google rely on exact word matching, *KartOO* explored the contents of sites and visualized the results as links on the basis of context. Exhibiting a form of tool intelligence, *KartOO* allowed us to make narrative connections with the results of its information foraging.[3]

The Dutch/Belgian hacker duo, known collectively as JODI.org (comprised of Joan Heermskirk and Dirk Paesmans), did not use semantics, connections, or narrative to tell its story, but instead implemented an aesthetics of disruption into its browser art. At their website, it is almost impossible to determine meaning or intentionality in their early works. They are collaged flashes of code fragments and piecemeal, animated garbage. Are we witnesses to or interactors with the screen events? The machine code seems to have a mind of its own. The screen flashes, windows open and close, and illegal downloads start of their own accord. In their works from 1995 to 2002, images of obsolete technologies depicted in ASCII march across the screen; since 2002, they have partaken of screen capture techniques which retransmit modified old games (including *Wolfenstein 3D* and *Max Payne 2*) to create nonsensical, deconstructivist, new game assemblages. In some ways, JODI.org's software really is the machine speaking as it hijacks our browsers and dances a ballet of indigestible code. To those who can read the code, JODI.org is seen to hunt down coding errors and bugs in software, pinpointing the sites of delay, shimmering screens, lurching movements, and endless feedback loops to expose a gallery of programming errors in other people's software (Kluitenberg, 2002) or give us instructions on how to build bombs (*Wikipedia*). "These so-called 'mistakes' then become not the disruption of a code, but the essence of the new code that jodi.org replaces the conventional ones with. In short what JODI.org creates is a set of negative signs or inscrutable messages that point towards the infinity of alternative codes of writing and reading networked media" (Kluitenberg, 2002). For the reader, JODI.org is cryptic in the extreme and, while there is clearly an intelligence at work here, it is so far beyond the human as to be incomprehensible.

The Impermanence Agent, developed in the U.S.A. by Noah Wardrip-Fruin, Adam Chapman, Brion Moss, and Duane Whitehurst from 1998 to 2002, used the contents of a person's Web browsings to tell the agent's own unique story. Agents are probably most familiar in the guise of the popular electronic pets, such as Tamagotchis and Furbys, and the Spirit and Opportunity rovers used by NASA to explore the surface of Mars. A more terrestrial and domesticated form of autonomous agent is the automated search engine. An electronic agent is defined by pioneer interface engineer Alan Kay as a goal-oriented system that can perform complex tasks, and that, most importantly, could ask questions and understand answers *in the vernacular* when it became confused or lost (Wardrip-Fruin). An agent is therefore an intelligent tool. German cultural critic Phoebe Sengers says that agents exhibit two particular types of intelligence: situated and cognitive (Sengers, 2011, 16). The Mars rovers, for instance, are situated agents, reacting as they do to specific problems in a particular environment. "Cognitive agents, on the other hand, engage in most of their activity at an abstract level and without reference to their concrete situation" (Sengers, 2011, 16). While *The Impermanence Agent* clearly falls into this latter category, as its name suggests it still does not conform to expected agent behavior. Instead, what its programmers set out to create was an anti-agent or an agent that you cannot interact with. Its mission is to interact with you (or, at least, with the contents of your browser) in order to tell stories.

Agents are somewhat eerie creations because they seem to demonstrate sentience. Sengers views agenthood as a metaphor for any machine that behaves like a living thing (Sengers, 14). The very process of naming a program an 'agent' means that it will demonstrate agency, that is the capacity for independent behaviour and for "engaging in complex action without human control" (Sengers, 10). An autonomous agent is a form of HCI. As such it is an interface designed for ease of use, but its user-friendly functioning means that it does not wait to be cued to respond.[4] It must be able to intuit a person's needs. An agent has attributes or "thought processes" that the designer instills in it and that inform its actions (Laurel, 1997, 144). For instance, Brian Lee Dae Yung Rowe describes his autonomous Digital Art(ist) *Daria* in terms of her 'motivations,' those being "to express herself and to communicate her ideas," and her 'goal,' that being to make a living from her art (users 'commission' works by making a donation through PayPal).[5] Sengers says that "If" agents are "'actually' some kind of tool, the creature is portrayed as fulfilling its … functions by being willing to do the user's bidding" (Sengers, 14). But what they most resemble is not a tool of any kind. It is an actor. As a performer playing a role, an agent's combined ability to learn and to react gives the impression of sentience behind its behavior. *Letizia,* for instance, an autonomous Web-browsing software, undertakes user profiling through keyword frequency to identify and display related pages that might be of interest. Brenda Laurel observes that agents increasingly have "traits that are *dynamic* (modified by learning and experience) and *relational* (modified in relation to objects and situations)" (Laurel, 1997, 146). These abilities are most familiar as human attributes. Being responsive, it is in an agent's ability to tell a story that engages a user's emotions, and thereby makes it seem real (Laurel, 1997, 147). That being said, Wardrip-Fruin's team contextualizes their project as being deliberately distinct from other kinds of agents. Their intention was for *The Agent* to function discursively as an agent, but to simultaneously question the role of agents while doing so (Wardrip-Fruin *et al.*, 2011). They were also conscious of the fact that they did not want this meta-agent to be a reactive engine or architecture that provided simply structural commentary on a user's surfing (Wardrip-Fruin *et al.* 2011).

The agent has three components that seem to mimic Bloom's domains of learning. First, it has a cognitive foundation – a fixed narrative text about impermanence and loss – and related images that are gradually washed away and replaced by sympathetic syntactical and visual structures encountered while browsing. Second, its physical attributes consist of a small window in which it is possible to watch the agent's progress as it sits in the background taking notes on and cutting and pasting from a user's preferences and travels over the course of a week. Finally, through the assumption of what appears to be an affective stance, it transforms itself into *The Agent*'s own story and accounting of a user's interests by rifling through the cache of a user's browser. One of its strangest attributes is that a user cannot interact with this piece directly. There is nowhere to point and click to reveal its contents; they are only visible when the user is browsing other websites. As the authors explain, *The Agent* extracts information from visited websites and adds its own as well, displaying the content unique to each user's foraging. Its contents return again and again, faded, altered, and adjusted, always running on the desktop whenever the browser window is open. Virus-like, *The Agent* steps in between your machine's http requests and the Web and delivers its catch, hauled in while surfing by proxy, to the browser. From there *The Agent* inserts images and text drawn from Web pages that are concerned specifically with its own themes – "impermanence, hypermedia, preservation, agency" (Wardrip-Fruin *et al.*, 2011) – on to other pages which the user has surfed. More importantly, *The Agent* also "writes back" inserting its own story into the user's movements over the course of five days: "They are *The Agent's* annotations, *The Agent's* mark, in the scrapbook of the user's experience. They are also a reminder to the user that they are working through *The Agent,* even when *The Agent*'s own window is not open" (Wardrip-Fruin *et al.*, 2011). Adam Chapman explains:

> we were endeavoring to make a work of art, so while The Agent had to be unobtrusive, it also had to be beautiful. And its beauty, its aesthetic engagement, was to be of a sort I would later hear the artist Camille Utterback refer to as *ambient interactivity.* The idea of ambient interactivity is simple: the work functions and is aesthetically pleasing on its own without the need for any direct interaction, however, the User is rewarded with an experience which becomes richer in direct relation to their level of interaction with the work. Thus, the work had to be aesthetically engaging in and of itself (Chapman, "A Well-Dressed Agent," in Wardrip-Fruin *et al.*, 2011).

All of this demonstrates that the story the machine tells is more complex than just its textual component. Just as the program slices text into grammatical segments, it divides its own images – all animated gifs – into layers, some of which get washed away and fade in and out. Both frames flip between text and image, with the left frame being the site of *The Agent's* voice. In a move evocative of the early agent and smart bot psychologist *Eliza, The Agent* counsels the user and consoles her on her 'loss' and walks the user "through the Kubler-Ross stages of grief as 404 'not found' errors are encountered during Web browsing (e.g., 'It must just be a typo. <pagename> can't really be gone'" (Wardrip-Fruin *et al.*, 2011). The 404 page names thereby become characters who are remembered, mourned, and their death denied. There is a poignancy to this that seeks to convince us that the agent feels emotions. This could be seen to tie in with Philip Agre and David Chapman's reconceptualization of agent subjectivity as it being more useful to see an agent as immersed in the events it is narrating rather than directing them (Sengers, 37). "An embodied agent," they say, "must lead a life, not solve problems" (qtd in Sengers, 37).

The Agent's right frame is markedly different. It contains *The Agent's* story. It holds images, stories, and fictions, particularly sepia-toned family photos and cemetery statuary. As the week proceeds, more and more of the linearity in the original text is leached away, and a collage-like narrative takes its place. While this all seems startlingly human, in fact what *The Agent* does is use metatagging to select less frequently used words, to give preference to nouns, and to rank words in terms of importance. By this means, "a subject of the user's browsing then becomes a subject of the story" (Wardrip-Fruin *et al.*, 2011). Much of the effectiveness of agents lies in this pre-programmed ability to remix language, syntax, and diction in distinctive ways. This raises the question whether this ability is actually a manifestation of intelligence. *The Agent* is distinctly different from autonomous AI engines. Its developers called it an 'opinionated archivist,' and one who customizes content (for instance, ads decay while other content lingers) using cookies, a proxy, and an IP address. Archivist-with-an-attitude is perhaps a clearer description of its function.

Where lies the work of art in such a mix? *The Impermanence Agent* developers have tried to demonstrate the work at galleries and the Guggenheim Museum, but time and space are integral components that inform the work. *The Agent* is with you over the course of a week, making itself at home like an out-of-town guest. It sits on your desktop, observing you, and slowly transforms itself in response to your interests and meanderings. What the developers discovered was that to exhibit such a piece was an act that did not translate well in non-electronic venues. The compelling quality it exhibits is in the way it engages with the user and in its act of creation. As a result, what Wardrip-Fruin *et al.* started to do was to present their documentation on the work itself: "the documentation," he says, "actually became the work itself; that is, the documentation is the artifact exhibited. In this way the exhibited *Agent* is moving closer to conceptual art than to process, performance, or artifact-based artwork" (Wardrip-Fruin *et al.*, 2011). Essential to this work is this very concept of impermanence. Not surprisingly, the work is unavailable to the current generation of browsers and the authors have already created several versions since they first started the project in 1998. This is one of the most pressing problems for the new media arts as a mode of expression. Obsolescence is the norm and that too is one of the things which makes *The Impermanence Agent* compelling.

We are on the verge of what Stewart Brand calls "a digital dark age" where data migration is required (at least) every ten years to survive the onslaught of new generations of software and platforms that make older information unreadable. As Jeff Rothenberg says, "Digital information lasts forever, or five years, whichever comes first." The end result, Brand predicts, will be a catastrophic loss of digital data. Everything informational is now stored, but will we be able to access this information and new generations of art when we need them? Among the biggest casualties of the disappearing data disaster are the new media arts. They are in fact more vulnerable than most forms because they are born-digital and can exist in no other medium. In addition, the very impossibility of archiving them informs their preoccupations and changes their act of creation. In short, the unarchivability of these works is not a bug but a feature.

Wardrip-Fruin and company are very aware of this fact. Decay and forget-fulness are the nature of the medium. The text rots away, parts of it falling into disuse from encroaching broken connections or missing files. As it decays it leaves behind only our memory of the embodied experience of our journey. It forces us to re-evaluate the medium. Instead of this loss being smart art's weakness, hubris, or failing, we need to recognize the way in which digital space foregrounds embodied experience in the present moment as its strength. This is what makes these bots, engines, and agents seem real. They are ethereal performers – like us. They are material, idiosyncratic, and subject to failing links, to memory loss. They seem mortal as a result of the insistent presence of their distinctive personalities and as a result of the particular performance they elicit from us. Because of this, they foreground notions of a new kind of embodiment. The new emergent model of performance in a digital age does not threaten old concepts of performance; it redefines them. What we see here is a paradigm shift in the relationships between performer, audience, and stage. As users, we are now drawn in to become the performer on the stage of the interface, and the audience has become this responsive little agent that acts on multiple sensory levels. Our bodies become multimodal as we navigate the immersive spaces of interactive art. Multimedia, by design, invites us to engage with it on a sensory level. It does this by emulating a whole spectrum of sensory interfaces. By collapsing all modalities into one continuous state, the environment becomes an extension of our skin and fosters a sensory effect that is akin to synaesthesia. This foregrounding of the sensory or experiential is performed on us as the agent engages us at the level of short-term memory or embodied experience. While the agent has the capacity for long-term memory, as an electronic archivist it is doomed to disappoint. This has been an ongoing theme in new media art from Janet Cohen, Keith Frank, and Jon Ippolito's *The Unreliable Archivist* to Marjorie Coverley Luesebrink's *Califia*. As an audience for our hopes, fears, and obsessions, the built-in obsolescence of these agents is able to mirror our concerns convincingly. As *The Impermanence Agent* performs its tale in response to our gestures, it invites us to engage anew and perform with it. Other search agents can harvest personal data and seem to be reactive to us.

The Japanese indie band SOUR, for instance, wanted to create a custom video for each viewer. They set out to develop an interactive app, called SOUR Mirror, that gathers data on its users to "transform your desktop into a stage" by creating a dancing figure, composed of personal information. One of the band's songs is about how everyone around us is a mirror reflecting us back; from that song, they wanted to make the Web a mirror. SOUR wanted to develop an app that would harvest social media data or Web-based info on an individual, so that the app could 'reflect' that person back to them. They raised the $5000 they needed for development through Kickstarter, and then invited their fans to set up Twitter accounts. By following the band's Twitter-feed, each fan's twitter icon becomes part of the official video for the band. When you log into the app's site (http://sour-mirror.jp/), you are are prompted to connect to your Twitter, Facebook, or Webcam. First, you see the app searching Google for images and extracting information related to you. It renders this data as a walking figure. Song lyrics are tweeting on your account, and the app harvests your locative data on Google Maps and calls up the map of your neighborhood (or your Internet Provider's neighbourhood). Your figure moves into a YouTube clip and then summons up your Facebook page. That page separates into nine windows that spatialize the image of each of the performing band members in turn. The grand finale projects your image on screen through your Webcam and you perform the finish as pixilated icons from your Twitter stream. Your pixilated image and video are downloadable and sharable. The site received 150,000 hits in the first three days and, unlike their usual home-grown audience, two-thirds of the viewers were outside of Japan. The experience itself is ephemeral. Apart from the downloadable component, none of the experience can be archived or preserved. In fact, as your data change, so does the experience of the song app. The Canadian pop band Arcade Fire's interactive film *The Wilderness Downtown* (2010) by Chris Milk also sets out to create a personal, sharable experience. Initially, viewers are prompted to enter the address of their childhood home. The video then constructs an experience around Google Maps' streetview information and aerial data. The end result is that the jogger in the video is running through your childhood neighborhood. Ultimately the video prompts you to enter some words that you would say to your younger self, if you were able. The old neighborhood is washed away and the results are lost.

The digital is vulnerable, fragile, and easily damaged. It has an extremely short lifespan, but it is that brevity, that breath, which imbues it with the gift of simultaneity – just as it is the performance of apparent sentience by these agents that makes us aware of their unnerving liveness. A performance cannot be housed or contained within the box of its medium – the computer in this case. Performance exceeds the medium's parameters. These agents reach outside their boxes to interact with us, making the text a responsive audience for our actions. If we do not engage, there is no performance. Meaning is born of the interaction itself. The importance of machine narrative and other digital artworks does not lie in a celebration of smart art, tool virtuosity, or greater freedom for the reader, but in the new windows it opens for the potentialities of embodied or multimodal reading and/or writing as a performative act. Dependent as we are on language, this shimmering art form foregrounds interaction (rather than conversation) and reminds us of those very human qualities we share with its opinionated archivists: fragility and impermanence. Does that mean that the agent is alive? Clearly not, but the seeming liveness of the medium endows the agent with what appears to be agency. Its liveness therefore is not so much a trick as a condition of its experiential or existential interface with us. It decays and fades away, but the experience lives on in our own storage medium – our memories, which, as in Wang Jianwei's reconceptualization of *Is He A Traitor? Post Production* (2002), sharpen into and out of focus over time.

Real time

On November 6, 2001, a new television program called *24* premiered; it was the first time that a television program had sought to document art as a real-time experience. ('Real time' is not to be confused with reality TV, which re/presents continuous events within a set timeframe; 'real time' foregrounds the temporal as a spatialized player and a part of the experience of the event.) That same week was the 50th anniversary of the comic classic *I Love Lucy*, the filmed live television show that invented the concept of the rerun and introduced the flashback to television for the first time (Hartigan, 2001, 16). Filmed on film, rather than taped to be taped over, Lucille Ball and Desi Arnaz negotiated for ownership of the shows on the assumption that people would want to see the episodes again. This had been an inconceivable concept up until that time. In a mere 50 years, we moved from live time to real time. Billed as "the most intense hour on television," *24*'s 24 episodes each spanned an hour in the course of one single day. Lucille Ball's laughs recorded a half hour in front of a live audience – one take, no second tries. *24*'s reality slowed time down, spatialized it, seizing 1440 minutes – minus commercials – of high action drama that played out in fast jump cuts and split screens of intervallic space and visual moments. Rather than having one case per show as is the norm in televised dramas, *24* introduced the concept of one case per season. It packed each minute full. At multiple points in the show too, we are held captive – transfixed – as the screen freeze frames not a single image, but on multiple moving images sharing screen space. They encircle a clock and are linked by a particular simultaneous sequence of time. This is a whole new rendition of the freeze frame from its traditional situation as a moment out of time to a new incarnation as a space in time. Fast forward 10 more years in the future to 2011, and the first real-time digital video wins the Golden Lion at the 54th Venice Biennale for Best Artist for

Christian Marclay at the ILLUMInations Exhibition for the 24-hour remix work, "The Clock." Immediately hailed as a masterpiece, the film remixes clips from thousands of movies in order to depict synchronized real time on the screen with the audience's viewing time for a 24-hour period. Searching for the footage, assembling, and editing the film took two years and an encyclopedic knowledge of film.

In *Welcome to the Desert of the Real*, philosopher Slavoj Zizek (2002) says that the drive for the real always results in spectacle, the opposite of real time, because time can only be captured as it passes, never in the present. According to Jordan Crandall,

> the real is only able to be sustained if we fictionalize it. To look for the real, then, is not to look for it directly: it is to look to our fictions, discerning how reality is 'transfunctionalized' through them. Perhaps the real object of the precision-drive is not only arrived at through reduction, but through expansion. To look to the object of the precision-drive is not only to narrow the optic, honing in on the target of attention: it is to look to the cultural fictions in which the object becomes lodged. It is to open the optic; theatricalize it. To accommodate cultural fictions is to acknowledge the constitutive role of conflict (Crandall, 2006).

The significance of this concept of real time should not be underestimated. It is a profound side effect of the fundamental shift which film and photography have made as a result of digitization. Paul Virilio aligns this shift in the visualization of the temporal with the magnitude of the discovery of real space perspective by Italian artists in the Renaissance. Real time, he argues, will begin to supersede real space, "making both distances and surfaces irrelevant in favor of the time-span, and an extremely short time-span at that" (Virilio, 1995). This is the domain of the interval: spatialized time or temporalized space. Where photography first made time visible by splitting the second into motionless water droplets or by revealing the horse airborne as it galloped, it is the shift from film to digital technologies that is making time with depth or spatialized time our primary mode of apprehension. It is also a time when we can look at old technologies anew. Sociologist Christopher Jenks has observed that vision is both a social and a cultural process, and technology – from microscopes to telescopes, spectacles to specula, cameras to computers – has long helped mediate the way we see. They are all useful tools for thinking about how we navigate the interlocking topologies of text and image in contemporary works of digital art.

Real time and time distortions are important modes of interpretation in remix culture. Wang Jianwei slows down – or speeds up – Vladimir Lenin's life in the Communist classic film *Lenin in 1918*. Short loops from the hugely popular film are played again and again in Wang's *Is He A Traitor? Post Production* (2002). In part, this shifts the focus on to the future role of Lenin becoming-politician. But as the movie becomes pixilated and degrades or sharpens and becomes clearer, the technical effects become a foreshadowing of the degradation of Soviet Communism. The sound desynchronizes too as the image degrades, and incorporates other appropriate music as mashups (*Synthetic Reality*). *Is He A Traitor?* was designed to be an exploration of the artist's own memory of the film, highlighting the parts that he remembers the best. The most famous real time work is probably Douglas Gordon's *24 Hour Psycho* (1993), a video work that slows down Hitchcock's masterpiece to a 24-hour duration. At the time it raised questions about the nature of authorship. In *Postproduction,* Bourriaud quotes him on the work:

"I'm still pretty skeptical about the concept of the author," says Douglas Gordon, "and I'm happy to remain in the background of a piece like *24 Hour Psycho* where Hitchcock is the dominant figure. Likewise, I share responsibility for Feature Film equally with the conductor James Conion and the musician Bernard Hermann … in appropriating extracts from film and music, we could say, actually, that we are creating time readymades, no longer out of daily objects but out of objects of part of our culture" (Bourriaud, 2002a, 86).

As Gordon puts on Hitchcock, Conion, and Hermann, we too slow down and watch and listen more intently. In 2008, he remixed his own work and produced a piece called *24 Hour Psycho Back and Forth and To and Fro.* Instead of the original single-channel long, slow play of the original film, this is a dual-channeled piece in which one channel plays forward from the opening, and the other channel plays backward from the end until they meet in the middle at the tortuous shower scene.

Technology can intervene in more than just the time of the playback too. In Chen Shaoxiong's digital video installation *Windows 2002* (a series he started with hand-drawn animation and photographs in 1996), buildings on Beijing's skyline dodge and duck and swerve in slow motion to avoid oncoming planes and missiles. Are the buildings swaying in time or in space? Is this a vision of the past or a premonition from the future? (This was part of the *Synthetic Reality* show in Beijing in 2004, the first digital media show in that country.)

Another technological alteration of time is Bill Spinhoven's *It's About Time/ The Time Stretcher,* developed and adapted by him from 1988 to 1994. The installation uses a closed-circuit slit video camera and monitor, which are manipulated by a networked computer. The computer shows the events onscreen 'live,' but temporally out of synch and elongated. Each movement the interactor makes is "rebuilt on the screen, permanently atomized and displaced." The user "dissolves in spirals, pixels, loops, and can, depending upon one's course of motion, completely dematerialize as if being beamed up by transporter" (Dressler, quoted in Golan, 2010). It is a disconcerting experience for the participant. Another eerie effect produced with the digital slit scan camera is the illusion of time flowing backward as a reflection of the image's stretched temporal duration. Spinhoven calls this "going back to the future." Jorge Luis Borges observed that "Time is forever dividing itself into innumerable futures," but he did not remark on how we stand at a fulcrum between future and past, and how our sense of time is wholly rooted in our sense of ourselves as embodied beings.

This is the area that Camille Utterback explores in a work called *Liquid Time* (2000). Concerned with embodiment and point of view, the interactor pushes deeper into time and reimagines the video frame as a plane of interaction. The curator describes the experience of interacting with the work:

> In the *Liquid Time Series* installation, a participant's physical motion in the installation space fragments time in a pre-recorded video clip. As the participant moves closer to the projection screen they push deeper into time – but only in the area of the screen directly in front of them. Beautiful and startling disruptions are created as people move through the installation space. As viewers move away, the fragmented image heals in their wake – like a pond returning to stillness. The interface of one's body – which can only exist in one place, at one time – becomes the means to create a space in which multiple times and perspectives coexist. The resulting imagery can be described as video cubism. To create this imagery Utterback's software deconstructs the video frame as the unit of playback (*Liquid Time Series*).

Like the 2D collages of the early modernist masters, Utterback's frame is a fluid one composed of the "sum of destructions," as Picasso dubbed the fragmentation of Cubism. As in the collage-like interface of Cubist works, Utterback's *Liquid Time* creates a space for the body and tries to portray every perspective at once by breaking temporal and spatial barriers. Like other slit scan technologies, this is an image that invites interactivity in multiple times and spaces simultaneously.

Ansen Seale also likens his work to Cubism. Similar to Golembewski's pinhole scan, Seale adapted a high-resolution panorama camera to shoot slit scan images on to a scanner. By disabling the 360-degree rotation mechanism, Seale has created a camera that shoots a scene one pixel width at a time in sequence. By shooting "a single sliver of space in rapid succession," Seale believes that his camera and scanner capture "a hidden reality" (*Ansen Seale Photography*). Inspired by the 1960s photographic methods of George Silk and William Larson, his 42-inch-x-100-foot-long backlit image called *River of Light* (2009) fixes the undulations of the water and transforms the still stones into stripes. He compares this work to a color field painting (*Ansen Seale Photography*):

> There is no Photoshop manipulation, says Seale. These distortions could really be described as a more accurate way of seeing the passage of time, even though it may be contrary to our traditional concepts of the depiction of time and space in art. In other words, my camera is recording a reality that exists, but one that we cannot see without it (quoted in Weiser, 2009).

Constant motion records the "'hidden reality' of movement through time" (Gillespie, 2007) as in his wildly transformed dancers. In the process of remixing reality, Seale adopts a painterly aesthetic in his work, ultimately being more interested in "the medium than the subject" (Gillespie, 2007). His chronoscopic approach is a new mobile vision that enables us to travel in and through a surrealistic imagescape (Gillespie, 2007).

GPS technology has taken us a long way toward realizing the implications of being able to animate and map information in space, but it is an art form known as locative media that is reconnecting us to our visual histories and fellow beings in urban spaces in the present. Digital media possess unique abilities to "transcend the boundaries of time, space and even language … to mediate historically produced ruptures that link past and present" (Faye Ginsberg, quoted in Meek, 2008, 21). This is an awareness of how the 'real' can be transformed, multiple histories, of oppressions, of the layerings of past, present and future generates irreconcilable differences and undermines any possible concept of "a singular history or theory of place" (Butt, 2008, 4). Mobile technologies raise the possibility, when combined with geospatial data, of making not just historical but mental landscapes visible. These ghostly presences are virtual histories, but "the virtual" is "a transformative dimension of the present" (Meek, 2008, 33). The slit scan interface meets mobile technologies in two iPhone apps:[1] *Timetracks* and *Foto Shaker*.

Foto Shaker by James Mattis (2009) is the simpler of the two. It has two functions. It has a twist mode, which distorts images, and shake mode, which blurs. By contrast, *Timetracks* (2010) allows for greater interactivity and control in the creative process of producing the final image. *Timetracks*, created by Masayuki Akamatsu, has five main actions: it will take One Way pictures, Palindrome pictures (which are the same forward and backward), and Still Scanned Moving images. It also allows you to adjust the width of the scan line and, just as importantly, the interval of the scan time. Needless to say, the images it produces are much more sophisticated than its twist-and-shake cousin.

Another dynamic imaging technology cousin that is relevant to the slit scan camera's capture of time and space is perspective. Perspective is a technology or tool for mapping an idealized relationship between our vision, our perception, and an object in the distance. Where traditional, linear perspective required a stationary viewer (as positioned by technologies like Alberti's window), digital perspective assumes a spectator who looks and is everywhere at once – both in space and in time. Digital perspective assumes a spectator both situated and in motion. Renaissance perspective made space visible through its rational use of geometry to define a fixed point of view. Painters, architects, and others constructed "an organized space, whose ordering principle – the central perspective – was seen as a guarantor of harmony and beauty in the presentation" (Jaschko, 2002, 1). Susanne Jaschko argues that fixed-point perspective mathematically encodes not only visible time, but space as well. Perspective makes a relationship between two subjects on a canvas apparent and measurable in "a mathematically-constructed space"; such a space adds "a time-related meaning, [since] a moment in movement is represented: the representation of the phases of movement, the freezing of a moment" within an action were designed to make the subject seem more realistic (Jaschko, 2002, 1).

Photographic film was able to not only depict a frozen moment for the eye, but to freeze it as it was literally in process. This gave it an air of authority or authenticity for depicting what Roland Barthes called "the scene itself, the literal reality" (Barthes, 2011, 17). André Bazin said that this fundamental aspect of photography's magical nature is "its capacity to engage with both the process and experience of time" (Metelerkamp, 2001, 5). Digital technologies open up all kinds of doors to spatial and temporal explorations that extend, complement, and cancel out the abilities of the analog experience. For instance, how do we classify time and space in a computer-generated work like Movie Bar Code? These visualizations of classic and award-winning feature films include every frame compressed to a thin slice, which are then lined up side by side. These striped panels become studies in aesthetics, mood, and duration, only freeze framed into a two-dimensional form. They become snapshots of the original film, making the whole of the film apprehendable in a single glance. The bar code engine takes the time-based original and renders it in a form that foregrounds the connoted message while the caption (which blends with our experience as spectators in the past) becomes the denoted message. Or does it? Does our remembered experience of these films themselves not in fact become a field of engagement for our study of the bar code version? Is this not a surrealistic snapshot of the place where our perception meets our imagination? Doesn't this version – for this is a whole new *reseeing* of the original film – constitute a poetic or aesthetic dimension of our experience of the film? This software translates the time of the film into the collapsed space of the digital image (about four feet long), and is reliant on the space of the viewers' imagination to reconstitute the original time-based work on the fly. At the same time, seeing the spatialized form of the color palate of the film as a whole may just reveal something new about the film and about our perceptions of it as well.

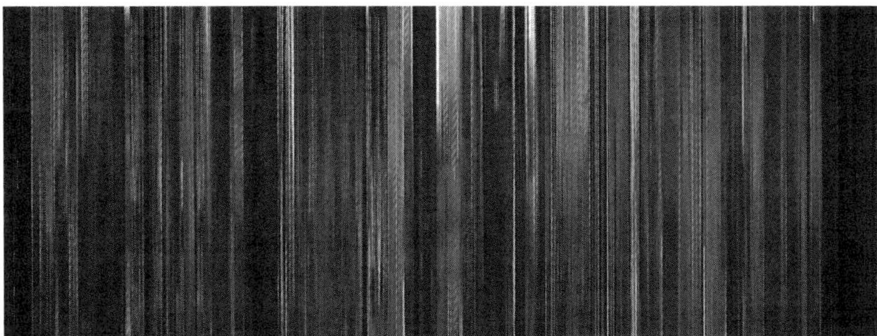

Figure 7 Movie Bar Code, *Vertigo*

Brendan Dawes' *Don't Look Now* (2004) sets out to accomplish a similar task to *Movie Bar Code*. It captures every single frame of the movie 1 pixel wide by 300 pixels in the slit scan-inspired processing java environment, and, yet, look at how strikingly different the effects are. Even though the resulting output is twisted into distorted shapes, says Dawes, "you can still make out parts of the actors and scenes. Tracking shots reveal themselves as stretched images while quick edits appear as staccato bars of colour" (Golan, 2010). Doesn't this further extend the suspenseful tones of the original film, drawing out the cadence of these scenes to an agonizing slowness? It also births monsters and aberrations, and the characters acquire multiple heads or stretched and distorted effects in their no longer human movements.

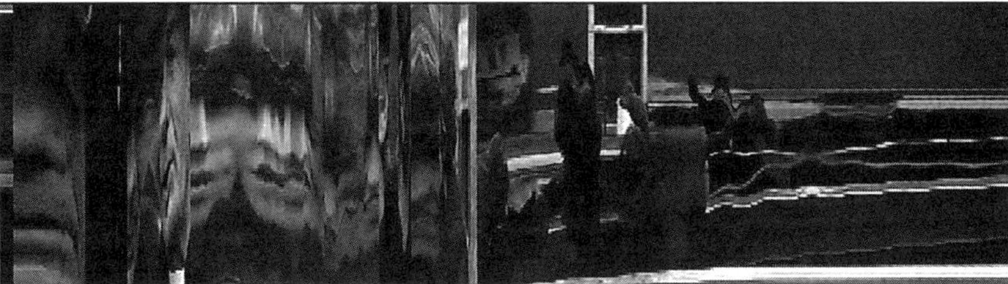

Figure 8 Brendan Dawes, *Don't Look Now*

PART THREE

Creative cannibalism and digital anthropophagy

Digital anthropophagy

In *Pacific Parables*, the Raqs Media Collective tells the story of the island of Nauru. It is a parable, they say, of the "toxicity of the gold rush," a fable of literally turning guano into gold at enormous environmental cost (Raqs, 2008, 14). They view this as a lesson in how the commodification of nature produces a toxic culture. For them, Nauru is an appropriate metaphor for how intellectual property law "takes for granted that the accumulated creative, imaginative and mental labor of our ancestors, which informs all our thought and creativity today, is a resource available for plunder" (Raqs, 2008, 15). They find real and telling parallels between the Guano Islands Act and the TRIPS Agreement. They say, "epics, stories, songs and sagas that represent in some way the collective heritage of humanity have survived only because the custodians took care not to lock them into a system of 'end usage', and embellish them, adding to their health and vitality, before passing them onto others" (Raqs, 2008, 15*)*. Under the Guano Island Act (August 18, 1856), any U.S. citizen can claim any island, rock, or key containing guano deposits anywhere in the world, providing they were not occupied or did not fall within the jurisdiction of another government. Similarly, the TRIPS Agreement (which came into effect on January 1, 1995 – and sets out the ground rules for IP in three areas: standards, enforcement, and dispute resolution) "allows citizens of several states to patent, trademark, copyright or otherwise assert their intellectual property claims on forms of life, aspects of knowledge systems, cultural materials and practices (wherever previous private intellectual property claims are absent)" (Raqs, 2008, 15). This, of course, overwrites and overrules differing conceptions of copyright, including the claims of, to name one group, aboriginal collective rights to their own tales and their traditional herbs for medicinal purposes. The TRIPS Agreement renders much human culture akin to the equivalent of guano atolls. "End-user agreements" are a "dungeon that prohibit cultural circulation and inhibit future creative activity" (Raqs, 2008, 15)."If we are to create cultural futures for ourselves, we will have to place our futures on the ground of a recovered continent" (Raqs, 2008, 18). That recovered continent of the future is the realm of the remix and DIY culture.

Creative practice is always instantiated in a place and a time and a people. Intellectual property has a life cycle that does not easily translate to an ownership economy. Economics have "almost completely neglected the question of" life cycles of property "in normal production and consumption"; A "'suppression of history' was endemic in market-based property systems," which only care about the now (Bagchi *et al.*, 2995, 5) and for far too long IP rights have been based on "affirmative protection for creative appropriation" – which sets the public domain in opposition to property rather than in concert with it (quoted in Sarai 4). It potentially gives corporate giants the freedom to steal from other creators. By contrast, digital poet Christopher Funkhouser and performance artist and found-footage filmmaker Vanessa Ramos-Velasquez both advocate for a form of cannibalism as the model for creative practice.

In "Digital Anthropophagy and The Anthropophagic Re-Manifesto for the Digital Age," Ramos-Velasquez calls for "A new practice of cultural consumption involving a technological mediation for input [both the feeding and the being fed], digestion and output." A Brazilian indigene, her culture has a long history of anthropophagic practices as a metaphor for creativity. She remixes the famous "Anthropophagic Manifesto" written by Oswald de Andrade in 1928, the leading Modernist poet in Brazil in his day, which "claimed that the Brazilian must 'devour' – critically assimilate rather than imitate – European codes using irreverence, inversion, joke, parody, sacrilege and insult as subversive anti-colonialist strategies" (Simanowski, 2010, 159). De Andrade's writings in turn were based on the history of Pero Afonso de Sardinha who in the mid-16th century sailed from Portugal to come to baptize the native Brazilians. The Aimorés tribe promptly ate him in a cannibalistic ritual. This is a foundational narrative in "the cultivation of heterogeneous expressive forms" for Brazilian creative practitioners (Funkhouser, 2007, 1.) De Andrade fell upon this history with gusto and proudly proclaimed, "'I am only interested in that which is not my own'" (quoted in Funkhouser, 2007, 1). Anthropophagy, or ritualized cannibalism, is the usual metaphor for creative practice in Brazil. Ramos-Velasquez translates this for the digital age. Under globalization, homogenization threatens, and she wants to retain the flavors of cultural uniqueness. She says,

> the allure, the attraction of "the other" is mutual and that it serves to
> form a symbiotic relationship that feeds both peoples. The concept of

"the exotic" is a two-way road, for if one has never seen the other before, their mutual discovery is of equal impact, and a curiosity to consume that newfound exoticism is occurring on both sides (Ramoz-Velasquez, 2011, "Re-Manifesto").

'Plunderphonics' (a term coined by Canadian composer John Oswald) has for far too long been the model of cultural use as the First World preys on the Third. Exoticism has been a facilitator, Ramos-Velasquez argues, in ensuring that that flow of goods was a one-way street. Now,

> However, because of the current high level of dissemination of information leading to a global democratic transparent society, that exoticism has become a multi-way road. And according to the dismantling of previous cultural imperialism models, culture is no longer just served from the top down as in mass culture, but horizontally as a global internetworked democratic village (Ramos-Velasquez, 2011, 6).

Raw material needs to be "metabolized" or reappropriated so that it becomes a two-way engagement, an engagement that transformed both the colonized and the colonizer or, in this case, both new and existing content creators (as in the works of Ni Haifeng discussed earlier).

This trope of "ingestion and digestion" (Simanowski, 2010, 159) has been further metabolized by Ramos-Velasquez, who finds we are all colonizers on the digital frontier. She uses remixing as a means and model for transforming found footage and copyrighted materials. She says, "the allure, the attraction of 'the other' is mutual and … it serves to form a symbiotic relationship that feeds both peoples. […] the indigenous cannibalistic practice resonates to today's civilized society, materializing as cannibalistic remix culture spanning the entire world" (Ramos-Velasquez, 2010, "Re-Manifesto"). It is not only our art materials that transform us, since everything we touch we transform as well. This renders the act of remixing a transformative and empowering one which links us not just with other artists and their materials, but with our technologies. What Auslander calls "liveness" is a mediated event brought into existence by recording technologies. It is this live, cannibalistic perform-ative act that has moved to the front and center of contemporary arts in our digital time.

> Everything starts as a concept, it evolves into writing, the writing gives shape to a piece, the piece then acquires a life of its own, which can take many forms of presentation. The conceptual art thus evolves into publications, performances, net art, interactions, interventions, and films.

The impetus is to combine elements into new contexts, proposing new perspectives and instigating thought (Ramos-Velasquez, 2011, "Quiet Revolution").

We all catch the creative contagion, but, as Charles Bernstein says, "a way to deal with that which is external … by eating that which is outside, ingesting it so that it becomes a part of you, it ceases to be external. By digesting you absorb" (quoted in Funkhouser, 2007, 11). Cannabalism-as-crowdsourcing can be a modus operandi too, as in British writer Christine Wilks' *remixworx*, a website where members have made 500 multimedia remixes in a five-year period (since 2006); Funkhouser (2011) describes *remixworx*:

> Artistic works are presented, and re-makes (responses) are posted in comment fields (that reflect how the works evolve). Beyond the high quality of the works, the collaborative axis of *remixworx* is more than respectable, and the sheer variety of types of works (stylistically/aesthetically) – kinetic visual poems often combining text/animation/sound – embraced by the collective is remarkable.

As far as digital anthropophagy as practice goes, Christopher Funkhouser identifies three different kinds of anthropophagic creation at play in digital poetry: (1) transcreation; (2) direct incorporation; and (3) mechanical presentation of work (appropriation of coding languages, etc.) (Funkhouser, 2011, "de Campos"). Transcreation is a process and a processing gesture. Direct incorporation allows for transcoding, mode switching, and the inclusion of images and sounds. In the mechanical aspect of the work, new navigational, architectural, and semantic pathways are formed. By absorbing all three, the author, the work, and *the audience* are transformed (although I should note that *audience* no longer has primary importance because in the remix economy we are all creators and we watch/listen *in order to re-create*). "To transform, one must be transformed. Incorporating anthropophagic conditions into progressive creative schemes is not compulsory … [but] opens up new promise for the synthesis of discrepant cultures and expressive histories" (Funkhouser, 2011, 11–12). This is in fact a new model of authorship and it opens the sluice-gates to new channels of communication and new kinds of art.

The reigning queen of new modes of communicating is Australian netwurker Mez (Mary-Anne Breeze). In the 1990s Mez invented her own fragmented language called *mezangelle*. It creates literary texts (fictional, poetic, cultural commentary, and other genres that have no name) out of a recombinant stream of leftover symbols, letters, fragments of words, slang,

IRC short forms, source code, markup, expressions, test patterns, protocol codes, and ASCII. Her work has been exhibited at all of the major electronic art venues and she has won many awards. Dubbed the Shakespeare of the code poem, she has also been compared to a meeting of James Joyce and Larry Wall (linguist author of Perl programming). Florian Cramer says,

> The beauty of "mezangelle" is that it uses elements of programming language syntax as material, i.e. reflecting formal programming language without being one. For the reader of mez's "netwurks", it remains all the more an open question whether the "mezangelle" para-code of parentheses and wildcard characters only mimics programming languages or is, at least partially, the product of programmed text filtering (data][h][bleeding texts).

Mez's writing explores ambiguity as an in-between space that links documentation to protocol without ever setting foot fully in either camp. Her rules are, in her own words, "elastic": a "distinct-yet-reasonably-elastic set of prearranged variables [protocols] I adhere 2 regarding each individual mezangelled piece. I decide on these prior 2 the actual construction of an individual wurk: 4 instance, they may include code conventions with subtle [poetic]alterations with a socially/software relevant theme" (Mez, email interview). The poetry is in the way the rules loop and remix within the set parameters, and play with the built-in protocols of the netwurk. According to Alexander Galloway in *Protocol*, "protocol is an affective, aesthetic force that has control over 'life' itself in the digital environment" (Thacker, 2004, xv). Contradictions are hardwired into the system based on the Internet's structure with a "bi-level logic" (Thacker, 2004, xv). Protocol is

> based on a contradiction between two opposing machines: one machine radically distributes control in autonomous locales, the other machine focuses control onto rigidly defined hierarchies. The tension between the two machines – a dialectical tension – creates a hospitable climate for protocological control (Galloway, 2004, 8).

The opposing machines are TCP/IP (Internet Protocol Suite), which is horizontal, and Domain Naming System (DNS), which is vertical, Galloway says. The TCP/IP establishes the rules by which network nodes communicate. By establishing the infrastructure for P2P communication, DNS forces all data into a vertical system of organization. As a translator then, the DNS database translates an IP address into a URL. The importance of domain names was made dramatically clear during the 2011 protests in Egypt where the deletion of a domain name wiped out a nation. It is believed that access in Egypt was terminated by shutting down the DNS, leaving 93 percent of the networks inaccessible (*Domain Registration News*). In the domain of protocol,

> Understanding these two dynamics in the Internet means understanding the essential ambivalence in the way that power functions in control societies. As *Protocol* states, "the founding principle of the Net is control – not freedom – control has existed from the beginning." To grasp "protocol" is to grasp the technical and the political dynamics of TCP/IP and DNS at the same time (Thacker, 2004, xv).

This is the space of Mezian play, manipulation, and logic. It is always in tension with itself, always existing in the contradictory spaces of multilevel logics. By mastering the logistics of control, she finds the free play in the system.

Mezian aesthetics operate always just outside of language in the not-quite visual realm. Composed of ruptures, openings, and jumps, her words are always in the midst of delivering poetic shocks and re-envisionings. Playful disturbance is the process by which the reader floats through the dynamic and sensory experience of reading her art-theory melange. Mez says,

> electronic media is altering [or at least subtly shifting] social engagement towards incorporating the idea of disturbance as normative [ie constant interruptions in geophysical space via mobile devices etc]; I perceive it as indicative of monumental changes in terms of human communication generally and specifically in relation to artistic creation and other forms of entertainment. As the core of my contemporary practice involves online/networked experimentation, I do c a type of pervasive collapsing of genres and forms; previous boundaries don't seem relevant in terms of some of the new artforms/communication pathways being produced. The resultant forms aren't even really classifiable as "art" or "entertainment" (Mez, interview).

Always pioneering new linguistic and performative spaces at the boundaries of form, in 2008 she was tweeter-in-online-residence for New Media Scotland. Her month-long project was called *Twitterwurking* and over those days she tweeted a complete, full-length work in mezangelle. The captured text was, of course, inherently leaky and so it spilled on to T-Shirts and other merchandise for participants. At the same time, she wrote a parallel "metacommentary" that ultimately took on a life of its own and became another discrete (if that is an appropriate description) poetic work.

Translation: Performing the in between

As soon as we move into the realm of language and in between spaces, it raises issues of translation. Canadian translator and theorist Barbara Godard aligns the act of translation with just such a performative movement in the present-tense rupture created by the "in-between" voices of counter-cultural artists. Artists inhabit a gap between languages, acting as translators of sorts who through their art make visible dominant cultural codes, expectations, language(s), and desires. Gillian Wearing, Eija-Liisa Ahtila, and Wang Jianwei have navigated this space in gallery installations. The countercultural translator literally performs this gap – the borderlands – between languages; she passes through the membrane and negotiates the space between as a process or a bi-directional gesture, like the act of remembering. This split has been a particular onerous one for Eastern artists living in the West. As Chen Zhen says, "living in the West as a Chinese artist is like having two cultures, two libraries, two armies" (quoted in Brouwer, 2004, 4). For far too long the burden of translation has fallen upon non-Western artists to make their visual works 'comprehensible' in the iconographic lingua franca of the West. (As Eastern art becomes more and more global, a creeping xenophobia has taken hold in a post 9/11 world accompanied by the greater restriction on immigration in the West, the increased use and acceptability of surveillance technologies, and the closing of borders.)

Where transcoding (what linguists call 'code-switching,' moving back and forth between tongues) grapples with the lack of equivalence between languages, and the disparity of their different modes of speaking, McLuhan struggled to remediate these issues in embodied form. Translation, and what cannot be translated, is always already about navigating between subjec-tivities where momentary equivalence is the best alternative that can be achieved because we can never truly inhabit another's subjective position. This is what is at play in Wearing's transgendered, transgenerational children. Equivalence, like a borderland or a queer space in straight culture, is always a self-aware cacophony of competing voices inside and outside of the source and target subjects and texts. Remediation is a translation process that uses equivalence as a means to allow us to slip our personal vernaculars into the space in between – and that includes the space between image and text, fiction and theory, vision and speech. Roberto Simanowski (2010, 159) discusses remediation in the context of anthropophagy:

> It may sound bizarre to talk about the remix of the discrepant in terms of cannibalism when there is a well-established academic term to describe such inter-media relationships. Almost a decade ago, Jay David Bolter and Richard Grusin used the concept of *remediation* to describe "the representation of one medium in another," a process in which the formal logic of prior media is refashioned in new media and which Bolter and Grusin mark as "competition or rivalry between the new media and the old". Remediation does seem to be a suitable concept to discuss the interdependent and competitive relationship between media. However, the notion of cannibalism and anthropophagy allows us to link inter-media relationships to an instructive example of appreciating and appropriating the Other in post-colonial history, which I want to revisit here in order to explore the hidden correlations between old and new media in the society of the spectacle.

This gap in which the Other can insert herself is the crucible within which transformations – linguistic, literary, mediated, contextual, cultural, structural, political, material – can occur. Just as the text is a set of interconnected systems operating within a set of nested systems within systems (Bassnett, 1991, 77), so the transformative gesture is implicit in the gap between voices, bodies, texts, and interfaces. Nothing is sacred in the interface either. It can also become a subject for remix. But where these gestures or movements used to play out in language, we now see them play out in recombinant images and recombinant image-texts.

"Different media produce different readers, different reading environ-ments and different reading practices," Rita Raley observes (2001, 2). These are the poetics of the extreme interface. The body as interface, as a writing surface, a writing instrument, and a dynamic voice generates agency and/or writes itself into the system from the outside. Subjectivities are irreducible and emerge from our experiential contradictions of our body as medium in motion. The subject is a social network that arises in the act of looking as we speedily move and browse through embodied textual spaces. Performance is key, for the concept of the work of art as animate object is as important as the concept of the body navigating the complexities of time in performative space. Think, for instance, of Canadian artist and VJ Steve Gibson's live club installations that are a mass of sonic material with visual snippets from Soviet science fiction films or other strange and unidentifiable media fragments.

I wonder, for us, whether the medium and message have actually converged, as Negroponte believes? Have body and text collapsed into one and the same – see-er and sayer, performer and audience – just as the computer collapses all modalities into a single technological literacy? Must we be dancers to understand the dance? Or, as Haraway maintains, can the digital and exploratory media birth a multiplicity of messages and new linguistic modes that perform entirely new kinds of embodied communication? Consider, for instance, Canadian found-footage filmmaker Richard Ker in *Collage d'Hollywood* (Canada, 2003). An eight-minute work filmed on 35mm film, it explored the materiality of the film medium in a literal way. *Collage* is assembled from movie trailers found at a deserted drive-in cinema, and explores onscreen sex and violence. Interrogating the grammar of Hollywood cinema, Kerr altered the surface of the film, "exposing grain, peeling and refolding emulsion like skin, and reconfiguring visual and content and subtitles" (Cammaer, 2005, 29). Video artist Naim June Paik also questioned the materiality of media, where he deformed television images with magnets or had interactors play and play with loose strips of reel by reel tape in *Random Access*. In reference to *Good Morning, Mr Orwell* (1984) – which was broadcast from Centre Pompideau in Paris and the studios of WNET-TV in New York, and was a media art piece that opened a path for reality TV – Paik says the work was:

> A heterogeneous mix of Pop (Peter Gabriel, Laurie Anderson, Philip Glass, Urban Sax) and avant-garde (Joseph Beuys, Ben Vautier, John Cage, Maurizio Kagel) [that] was electronically collaged and transformed. Through a split-screen technique, the TV picture showed simultaneous events occurring in different locations. *Good Morning, Mr Orwell* was broadcast at the same time in Korea, the Netherlands, and Germany. The panoptic principle turned into the pleasure principle (Weibel, 2002, 6).

How many modalities and ways of seeing and hearing have been collapsed here?

Collaged works enact the aesthetics of interruption, disturbance, and capture/leakage in a variety of ways. They can either join extra material together as in Dayna McLeod's rendition of *Nothing Compares To You* by Gregory House (Hugh Laurie) or delete material to make subtexts more apparent as in McLeod's *Don't Ask Don't Tell Gay, Gay, Gay*. McLeod says of her work:

> Aesthetically and artistically, the collage videos are about the content and cutting to the chase. Recently, I've been working with both an additive and reductive process: *Nothing Compares to You* is additive – I'm searching for specific content and putting it together to make something new from the source material. *Don't Ask Don't Tell Gay, Gay, Gay* and the Secret Messages series are reductive because I'm taking the episode or film, and cutting out the irrelevant content (to my rules). After the challenge is met, I then have to make decisions about the final cut – adding the sweeping melodramatic chords from the song for *Nothing Compares to You*, the timing of the music, do I cover up the cuts for words I've spliced together, how do I end the piece, what happens in the 20 second bridge in the middle of the video? These are all choices that get resolved in the edit, but are obviously integral to the final version (Hogan, 2011).

Don't Ask Don't Tell Gay, Gay, Gay plays institutionalized prejudice for laughs by paring down an episode of the television show *Boston Legal* to the queer content in a court case that explores the 'Don't Ask, Don't Tell' policy of the American military system. McLeod's translation of mainstream media's social codes is in sympathy with Cao Fei's use of the traditional Chinese art form of shadow puppetry in combination with Communist iconography in *Shadow Life* (2011) to produce a scathing critique of totalitarianism that in no way resembles children's theater.[1]

'Productive mistranslation' (China and Pakistan)

Asia is a special case when it comes to its role in digital art and authorship – and not just because the region is the major source of pirated digital materials. Asia is a complex cacophony of cultures, languages, peoples, and regions. It is difficult to nail down particular trends even within a single country, let alone in the region as a whole. There are in fact many different Asias. That being said, there are individual artists and collectives who are emerging players. In the realm of digital art, they do not merely slavishly imitate Euro-American forms (as some would argue), but are practicing what art critic Tatehata Akira calls 'productive mistranslation' – a similar approach to digital anthropophagy. Moving beyond a simple reapplication of existing techniques, they are adapting and refining technologies and atactical methodologies to their own ends. I will focus on China and Pakistan in these pages, because they are unique cultures whose art has emerged or is emerging out of difficult political situations and bloody confrontations, and because they are the major centers for pirated digital materials. China, poor in intellectual property for many years, is undergoing a massive cultural renaissance. Pakistan, a culturally rich region with a long history of recognizable artistic practice, is in the midst of morphing, under strict censorship and the suppression of culture by Al-Qaeda and the Taliban, to find a way to apply its creative energies under the radar in activist undertakings that educate and protest the silencing of the cultural sphere.

It is precisely in the social process of reseeing, interrupting, disturbing and capturing, leaking, digesting and regurgitating culture – a process that the Taliban, for one, seek so hard to suppress – that the productive mistranslation can be born. It is in the shift from product to process that the antidote to and outside of consumer culture may be found. In *Postproduction,* Nicolas Bourriaud notes Jean-François Lyotard's horror at the slippage between the postmodern condition, which was an informed blending of styles. The cataclysm of "the so-called postmodernist art of the eighties: to mix neo- or hyperrealist motifs on the same surface with abstract, lyrical or conceptual motifs was to signify that everything was equal because it could be consumed" (2000a, 91–2). Lyotard's refined sensibilities, which favored only highbrow art, feared a descent into the distraction of advertising's modular design as in magazines or supermarkets – where things were merely designed to be consumed. In support of Lyotard, Bourriaud cites Yves-Alain Bois who also underlines the situated and contextualized history of forms as a net against a decline into cynicism and a catch-all for the collapse between the distinction of the highbrow and the lowbrow (2000a, 92). That collapse is upon us and perhaps more remarkably it has become the barrier that shields process-oriented contemporary art by suggesting alternate strategies to prevent it also from being subsumed into consumer culture.

Over the past decade in Asia, there has increasingly been what David Teh in "The Video Agenda in Southeast Asia, or, 'Digital, So Not Digital'" calls a "democraticizing effect" as access to technology – including computers, cheap pirated software and cameras – has spawned new art forms and new artists. What is largely absent in Asian cultures, however, is a *networked* aesthetic or networked art. Spaces, gaps, emerge instead for productive mistranslations. As we have seen elsewhere, digital artists in the 21st century are tending to favor "lo-fi over hi-fi" (Teh, 2011, 83); in addition, censorship (including self-censorship), and cultural or religious edicts may inhibit the use of images or discussion of particular topics (as in religious restrictions). Clearly political and technological stability (as in the constant electricity brownouts in India and Pakistan), as well as the cost of online access and a pay-per-use Internet service provider (ISP) model contribute to artists shying not so much from the digital medium as from online access in general and from the ever more increasingly walled or monitored World Wide Web. Where "Networked art … remains a fringe activity, barely visible in most of the region," Teh says that the use of networked communications continues to expand for promoting artists' work and for communication near and remote, and for "community building, research and inspiration, documentation and debate" (Teh, 2011, 163). The influence which networked communication has had on contemporary art, cultural, social, and economic practices is immeasurable – more important, he argues, "than technical, industrial or art histories" (Teh, 2011, 163).

Teh's main subjects of study are digital culture in Thailand and Indonesia, but his findings seem to be applicable to much of Asia. He notes that artist and activist modes often merge (Teh, 2011, 163), a finding that corresponds to my own research in Pakistan, as I will discuss later in this chapter. He says that "'new media art' is almost invisible" in the region (Teh, 2011, 163). He has found "only a few Thai artists working with new media in reflexive ways." Teh (2011, 168) continues:

Even the few who engage with network *thinking* seem compelled to 'realize' their ideas offline, in more tangible physical forms. Here, the pop-media projects of Wit Pimkanchanapong are exemplary. In these works, video, computer graphics and mobile and locative media converge not in virtual space, but in the very physical, social spaces of the shopping mall and the rock festival. Wit is Thailand's most conspicuous new media artist, but even in his practice the web is only really a channel for promotion and documentation.

Teh does not consider the possibility that the documentation might actually *be* the work, as I have argued in these pages. The reality is though that artists, especially in developing countries, more often than not need *objects to sell* in order to make a living. The ephemeral nature of new media does not generally fit that bill very well. Some media artists have done well, like Cory Archangel and Feng Mengbo (who, for instance, sold his game *The Long March: Restart* to the Museum of Modern Art in New York), but they are the exceptions. Even in China where support for artists is significant, many continue to paint on the side for financial reasons. Another Thai artist, the conceptualist Pratchaya Phinthong, practices social sculpture. Social sculpture was an idea conceived by German conceptual artist Joseph Beuys who called for an extended definition of art. Beuys believed that the whole of society should be treated like a work of art: a *gesamtkunstwerk*, a total work of art (an idea that originated with Richard Wagner which led to the development of the modern-day concert hall). Teh says that Phinthong's social sculpture is rooted in "social and technical systems of knowledge dissemination" (Teh, 2011, 163). Despite this, Phinthong does not consider himself to be a new media artist:

In his 2006 installation *Alone Together, Pratchaya* created a quasi-domestic space in the gallery where visitors could watch cult and art movies from a DVD library, as well as duplicate films to take home or add to the library from their own collection – a kind of peer-to-peer network. Again, new media thinking (in this case concerning international film piracy and its local market nuances) gravitates towards shared, physical space, and the face-to-face encounter (Teh, 2011, 168).

Art *is* knowledge in spaces where there is restricted access to information, restrictive regimes, or low literacy rates. Peer-to-peer information sharing is not enforced or monitored in Asia and so there would be no restrictions on sharing digital information in this kind of venue any more than there is on sharing any other kind of information (yet). In China, pirate DVD stores live side by side with police stations. In fact, in Kunming province alone it was discovered in the summer of 2011 that there are nine fake Apple stores that sell real Apple products. Other fake Apple stores exist in Croatia and Venezuela. Those fake stores (re)sell real Apple products. If you want to buy fake Apple products (like the iPed in Japan, a fake iPad which runs both Android and iOS, or counterfeit Apple iPhones that are entirely re-engineered with similar appearance and functionality but less reliability), then you can do so in any piracy mall or pirate electronics retailer in Asia. As far as pirate DVDs go, they used to be of poor quality, were missing features or part of the film, or sported atrociously poor audio since they had been filmed illegally in theaters (a form known as cam-captures), but now most DVDs are virtually indistinguishable from the 'real' thing.[1] This is not surprising given the fact that the activity that media companies are really in the business of is selling *copies* of easily replicated digital data all day, every day.

Teh notes that, while his primary concern is with video production, there are collaborative structures of authorship already in existence in these cultures, for safety's sake. He says,

> in places where a politics is based on expression of individual will is neither native nor entrenched – and where the state has largely been and largely remains its antagonist rather than its guarantor – the concept of voice denotes something for more quotidian, variable, and shared. In such contexts, voice is just as likely to be a vehicle for the proliferation and scrambling of identity codes (Teh, 2011, 173)

This is so in new media groups like the artivist troop the Mauj Collective in Karachi, Pakistan, or the art-activist-musicians Taring Padi (Fangs of Rice) in Bantul, Indonesia. In addition, Asian cultures by and large are not rooted in the culture of the subject and in identity formations in the same way that Euro-American cultures are, and that egocentric 2.0 culture in particular revels in. It was modernity that was in love with the politics of voice and that "has invested voice with the politics of identity as something 'inalienable' belonging to the individual subject, gradually recognized and secured by the modern state" (Teh, 2011, 173).

That shift is starting to make itself felt in art in China where artists like Fang Lijun, Miao Xiaochun, Feng Mengbo, and Cao Fei insert themselves as a subject into their art. It is important to note that in a country that has no tradition of self-portraiture (and a slim at best one of portraiture), the use of the self as a subject is operating in very different ways than it does as a stock-in-trade in Euro-American notions of identity. And these artists take them to extremes: Cao's subject appears as her avatar in her digital city in Second Life, and participates and observes as tourist, observer, and guide, Miao's virtual environments are populated by identical versions of himself that number in the hundreds in virtual environments, and, in Feng's hack of the popular game Quake, *Q4U* (2002), he appears as both fighting self and camera-gun-wielding enemy to be exterminated. While these artists are clearly the 'authors' of their selves and their art, the very notion of interactivity in exploring China Tracy's RMB City, Miao's worlds derived from European paintings, or in playing Feng's game problematizes authorship in complex ways. More and more in global digital media, we find crowdsourced work or collaborative projects – like Cizek's *Out My Window: Highrise* or Bard's *Man with a Movie Camera* or *Bermain (Playing)*, a collaboratively produced video by ruangrupa, an artist-run space in Jakarta, and by the youth radio collective Prambors Radio (2007)[2] – complicating what it means to create. The "proliferation of channels, voices and tones" (Teh, 2011, 174) in response to an always-on world where there is no longer a private sphere, and to digital possibilities in multimodal expression leads to what Teh calls "tactical distraction or *encryption*, a diffraction of identity in keeping with an era of total digital diffusion" (Teh, 2011, 175, emphasis in original). Teh argues that "The *auteur's* response to the end of privacy is, perhaps paradoxically, the media equivalent of speaking in tongues, a multi-vocal channeling that undermines the unity of a single, authorial voice" (Teh, 2011, 175). Or the death of authorship as we have known it altogether.

The devolution of authorship is just one symptom of the proliferation of voices. Digital media also widen the range of tones available to the individual, as in the work of John Torres, an emerging digital author from the Philippines. Torres' images are remarkable for their unpolished candor. His unstaged, clearly non-industrial manner of production yields a kind of digital personal realism, distinct from the celluloid social realism that loom so large in the Philippines' film history. Torres' videos exemplify the economy of 'always-on' digital media, culled from a constant, unscripted recording of quotidian life. His DV camera sits within a wider ecology of convergent media – PCs, mobiles, Dictaphones and voicemail, radio.

Television, karaoke, and the web – which are both channeled and represented in his loose, poetic narratives. (Teh, 2011, 174).

Like Khaled Hafez's *The Video Diaries* (2011), these are poetic narratives that increasingly embrace or incorporate other people's stories. In politically dangerous times (and since we are always audible, trackable, and monitorable when are our times *not* dangerous anymore?), such kinds of first-person modes of expression can be potentially fatal. Witness the publishing of *Ai Weiwei's Blog* (Weiwei, 2006–9) in print in the United States in March of 2011, for instance, as a way of preserving documents that had been or were expected to be censored. Multiple formats, Teh notes, are a safe response to surveillance cultures; multiple modes and venues are harder to suppress. New models for collaborative art or relational architectures outside of the gallery likewise seem to have been inhibited for reasons of access as cited above. In Thailand, video seems instead to adhere more to an old-style journalism model and digital media art still conforms to the unique, identifiable author model. Teh (2011, 173) says that

> If the video agenda in Thailand conforms to what [Manuel] Castells called 'networked individualism' (a networked form, that is, of 'sociability'), the shapes it takes in Indonesia looks more like what Ned Rossiter calls 'organized networks' – they are transdisciplinary, distributed and collaborative.' This distinction is broadly confirmed by comparing authorial modes in each country.

It is also reinforced by codeswitching across modes of discourse (image + text), media, and languages.

In Pakistan and India, where the artist communities are small but internationally networked, and where English as the language of higher education has removed barriers to access for those lucky few, awareness of both greater artistic movements, practices, and materials is profound. In 2000, Huma Mulji of Pakistan and Shilpa Gupta of India came together to foster a conversation in the first of three cross-border art interventions. The *Aar-Paar Projects* took place in 2000, 2002, and 2004. They arose out of conversations which Mulji and Gupta had at a conference in New Delhi, India at a time when relations between Pakistan and India were just starting to reopen. The two thought that they might cut through the red tape, bureaucracy, and gallery structures to create a public art project on their own. Meaning 'through and through,' or 'this side and that side' to describe how a needle moves through fabric, *Aar-Paar* addressed the reality of Partition, divided lives and states, and issues of language. In sharp contrast to a linguistically divided China, for instance, where more than 90 regional languages were historically linked only by a common written language, Urdu and Hindi are derived from the same linguistic family roots and are intelligible to residents of both countries. Their written alphabets, however, are derived from their cultural divisions: written Urdu is based on its Persian origins and written Hindi on Sanskrit, a division that dates back to the 18th century. The 38-letter Urdu alphabet is written left-to-right and the 96-letter Hindi alphabet is written right-to-left. Hindustani (Hindi-Urdu) is the mutually intelligible, colloquial language of Indian popular culture that is spoken, for instance, in Bollywood films. *Aar-Paar* was also the title of a 1954 classic Bollywood film directed by and starring the legendary Guru Dutt, one of the most influential people in Hindi cinema.

In 2000, digital technology was not yet easily available to artists in the two countries and so the first *Aar-Paar* show was conceived as a mail art project. The second was fully wired, with the art distributed by email for printing in the other country, and the third was a video project that was shown in galleries. Conceived as a joint venture between Karachi and Mumbai artists to bridge the rift that followed Partition for cities that had a long and rich history, the first *Aar-Paar* project involved five artists from each country. Submissions were by invitation since there was no infrastructure to run the show as an open call or any other way. Their original submissions were sent via normal mail. Censorship between the two countries was a big problem, and, naturally, items with the stamp of the other country were more likely to be targeted so that some materials arrived damaged and some did not arrive at all. Mulji and Gupta intended this to be a public art project or, as Mulji says, "a physical peer-to-peer network" (interview); to achieve this, they spoke to people in markets to decide on the best locations for posting. They chose the Third World equivalent of news kiosks – roadside food stands and paan shops – places where people would go to buy a single cigarette or a paan to be consumed on the spot, or to advertise or search for work, etc. The organizers state, "The idea of intervening in the city was to create an alternative audience, extend viewership of art beyond an exclusive 'art' audience and to take people out of their comfort zones for art viewing in both cities" (*Aar-Paar Text*). While the art surely enriched the lives of the people in the poor neighborhoods, there is no reason to believe that it affected gallery-goers' usual behavior. As an art network, the art was posted for a 10-day period, creating an Internet of things and including an example of each work. Much of the material consisted of remixes.

Asma Mundrawala's *Eid Attraction* plays on the Bollywood tradition of action films having Hindi men woo, convert, and dominate Muslim women. The films integrate the Muslim heritage into Hindu culture and subjugate it. The reverse never occurs, except on Mundrawala's poster. She blends a Lollywood (Pakistani) hero with a Bollywood (Indian) heroine to create a new imaginary film, *Kabhi Haan, Kabhi Naan*, which translates as "Sometimes yes, sometimes no" (Duelling Partners). Crossed guns at the top of the image underline the inherent danger in a compromising position:

> The digital collage includes a burst of fire in the background, suggesting the volatility of this pairing, and its title echoes dangers present in fusing the two nations together. "Sometimes yes, sometimes no" refers to unstable politics and personal associations between the two countries – "sometimes friends, sometimes not" and "sometimes admirers, sometimes not." In this duality, the artist seems to suggest that even though the two countries might appear to be alike, they are always divided in politics; even as there is a shared culture, underlying mistrust is now part of any dealing with other countries in South Asia (Duelling Partners).

The coy expression on the woman's face, however, undercuts all seriousness and leaves no doubt that this is a playful send-up. The vacillating position the poster takes is clearly intended to capture the nature of a love affair where opposites attract – a favorite theme for comic and romantic films of all stripes.

The most controversial of the posters was one by Roohi Ahmed; it included a map of the region with question marks around the disputed territories, and the work "hum," which is a mashup of 'he' (the first letter of 'Hindi') plus 'mum' (the first letter of 'Muslim') meaning 'us.' Functioning as a tiny dictionary, the poster used Urdu words from the dictionary that have 'hum' as the first syllable. The resulting texts parses as "we are the same age, we are the same class, we are the same color," etc. The difficulty in posting it came with the map itself, not the content per se. Merchants refused to post it, claiming the map was "wrong." When the two official maps of the countries were compared, they were indeed different. India, for instance, allowed no disputed territory. The intention of the poster was to underline the peoples' *similarities* instead of their *differences*, the latter being where the focus in disputes usually falls.

By the time of the second show in 2002, Pakistani artists had gained access to digital technology and were wired to the Web. This was fortunate, since, in the wake of a military clash in Gujarat where 1000 people had died, relations between the two countries had deteriorated considerably with escalating threats in a nuclear arms race. *Aar-Paar* as a result was launched in a much more aggressive display of guerrilla art by 10 artists from each country. Twenty posters were printed in runs of 1000 each – with each artist contributing a small sum to offset printing costs. Those 20,000 prints were plastered all over Mumbai and Karachi. Illegally posted as a part of their artists' statement, this artistic intervention was designed to provoke thought at street level. Posters were pasted in multiples (in the manner of advertising) so that they

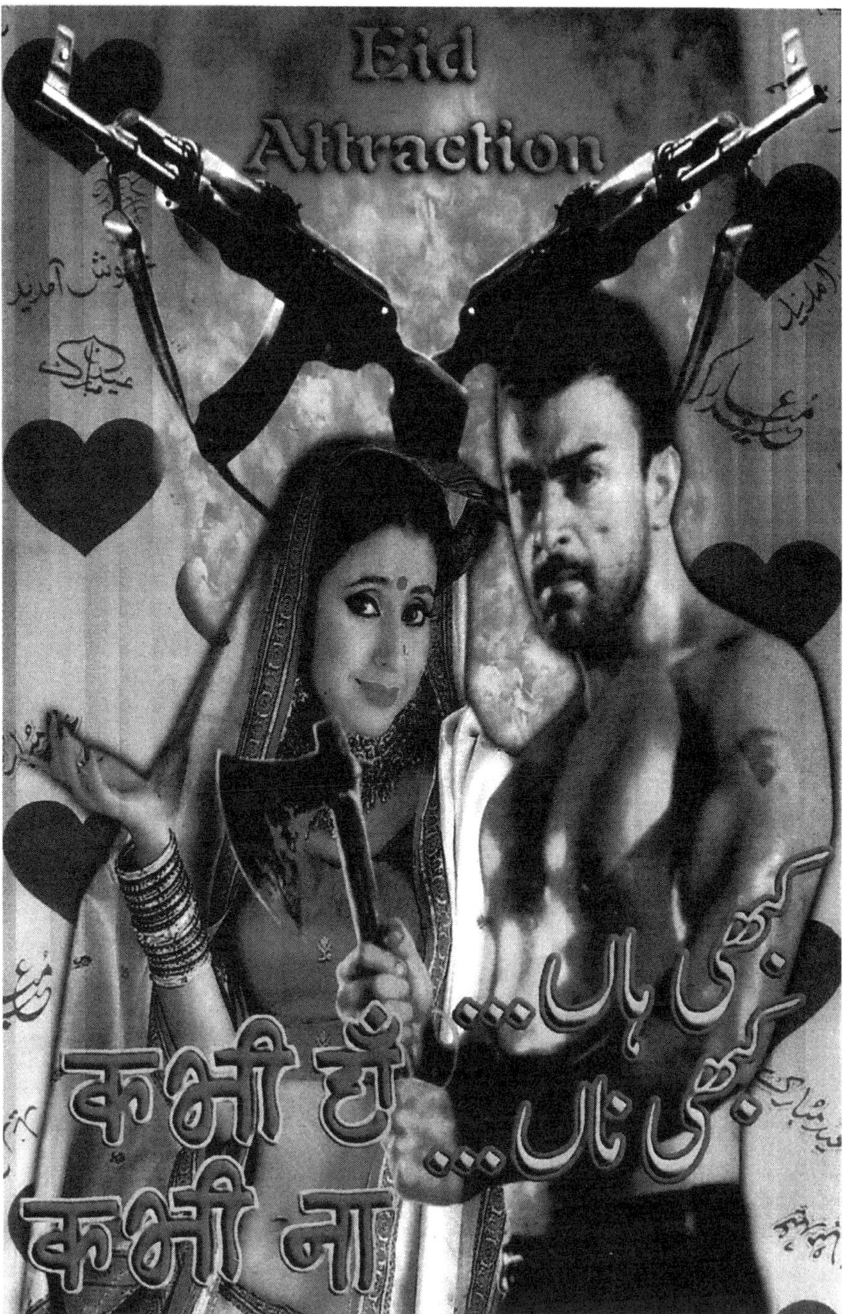

Figure 9 Asma Mundrawala, *Eid Attraction* (2000)

could not be missed. These posters were even more visually arresting than the first. Roohi Ahmed's *Karachi: A Culinary Map*, Shez Dawood's *Together We Are Stronger*,[3] and Gupta's *Blame* bear a startling resemblance to the Situationist International posters from the May 1968 intervention. Ahmed's map was locative media for restaurants with Indian cities in their names. Its purpose was to draw parallels in the relationships between the two countries' cuisines. Even so, Gupta was almost arrested again (she had been detained for two days on account of Ahmed's first map) because his poster violated a linguistic edict: it is forbidden to display Urdu in public in India. Dawood's poster, *Together We Are Stronger,* is a deceptively simple comment on the arms race, which depicts two nuclear warheads, each emblazened with the disputing countries' insignia. The underlying message advocates in favor of the power of cooperation over the might of conflict. Gupta's poster, *Blame,* was red and white, with an image of a bottle with the label "Blame." The text beneath the bottle reads (in both Urdu and English):

BLAMING YOU MAKES ME FEEL SO GOOD
So I blame you for what You Cannot Control
YOUR RELIGION YOUR NATIONALITY
I WANT TO BLAME YOU. IT MAKES ME FEEL GOOD.

Gupta also handed out small bottles with the label that were filled with simulated blood. By highlighting the bloodshed that arises from irrational personal or public sentiment, Gupta cuts to the heart of the problem in sectarian conflict.

Another artist who participated is the greatest Pakistani remix artist, Rashid Rana. In his controversial and disturbing *Veil Series* I, II, and III (2004), he draws upon the Pakistani miniature tradition to digitally construct pictures of burqa-ed women out of thousands of images of naked, fornicating bodies from hardcore porn sites. In *The World is Not Enough* (2006), he stitches together an abstract 'aerial' image of immense beauty from dozens of his own photographs of landfill sites. For *Aar-Paar,* he produced *In the Middle of Nowhere.* It is a bi-cultural dual image of himself photographed as a pieta with each figure in the garb of one of the countries, set against a romanticized painted backdrop of a Kashmir landscape. In this area known for its sublime scenery, Rana poses holding his dead self in his arms as he kneels in a painted river. As a (further) critique of multinationals, he has inserted a gas station into the middle distance of the idyllic scene. Finally, artist Jitish Kallat contributed a Google piece. His poster includes two screens from a Google search. The first displays a search for the word 'peace.' The second shows the results: "Your search – PEACE – did not match any documents."[4] Suggestions are offered as to how to refine the search with alternate spellings.

The third and final show in 2004 was of video works, which were shown in galleries in Pakistan and elsewhere. In the intervening years, India has changed so much, and Pakistan has been so transformed by the arrival of the Taliban that it is unlikely such an effort could be attempted or even desirable in the foreseeable future. Chantal Mouffe sees public art in social spaces as a necessary condition for the resistance of spectacle and capitalism:

> What is needed is widening the field of artistic intervention, by intervening directly in a multiplicity of social spaces in order to oppose the program of total social mobilization of capitalism. The objective should be to undermine the imaginary environment necessary for its reproduction. As Brian Holmes puts it, 'Art can offer a chance for society to collectively reflect on the imaginary figures it depends upon for its very consistency, its self-understanding' (Mouffe, 2007).

In the case of Pakistan and India, such self-understanding also involves coming to heal the traumatic wound that Partition inflicted. Art activism helped bring this issue to the fore and provide a bridge in a time when censorship would have prevented these overtures from being made in any other way. Art in the streets on a shoestring budget helped spark new collaborative organizations

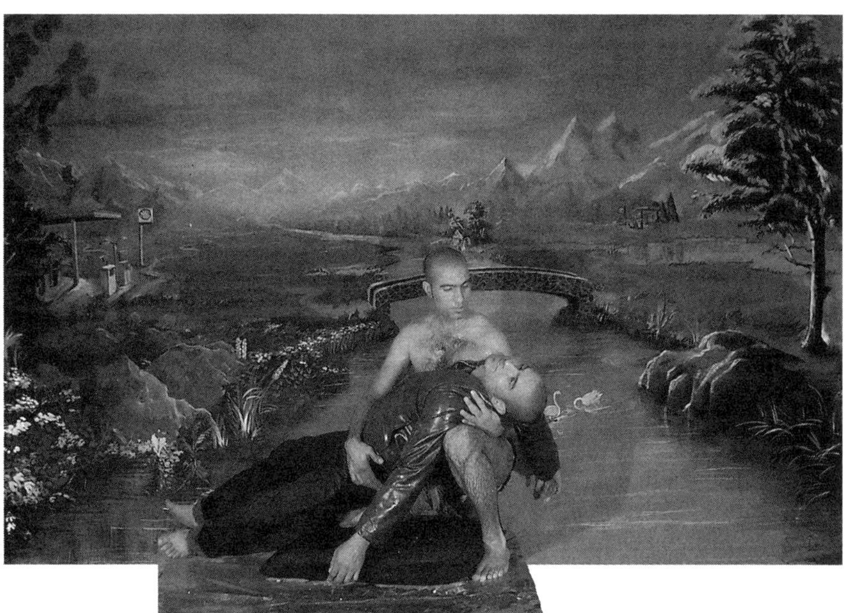

Figure 10 Rashid Rana, *In the Middle of Nowhere* (2002)

and expressions of contemporary art. Unfortunately, such initiatives are now impossible in a time when Al-Qaeda are on the loose, whereas it made a big difference then.

Another group initiative in Pakistan is the Mauj Collective in Karachi. Arising out of conversations by a local group of artists from February to June 2007, it initially had four members: Nameera Ahmed, Amar Mahboob, Yasir Husain, and Atteqa Malik. Working as facilitators, interventionists, educators, and artivists, they set out to make technology available without cost to anyone who wanted access and to advocate for free and open source software. Only once the technology was available to people at large could they begin to connect technology and culture in creative projects. All were accomplished artists, and, in addition, Husain had been working with a group called Hill Park to try to open a cultural center, and Mahboob had a Linux background, so with this combination of skills and interests the group congealed quickly. Malik is a tireless organizer of events and grant proposal writer, and did much of the Collective's promotion on top of her own art (which has been featured at the Tehran Biennale and elsewhere). Mauj became very successful very fast and quickly grew to an ensemble with 13 members. They would present and teach at fairs and public events, teach classes, and travel around the world to talk about the initiatives they had underway on everything from global warming to media art activism. Over a period of several years, they introduced hundreds of schoolchildren to digital technologies. They appeared at the Goethe Institute in Karachi, at Transmediale in Berlin, at the Civil Society Art and New Summit in Indonesia, and at Ars Electronica in Austria. On November 1st, 2010, the show *The Rising Tide : New Directions in Art from Pakistan 1990–2010*[5] opened at the Mohatta Palace in Karachi. It was a landmark contemporary art show featuring the work of 42 of the country's leading artists, both emerging and established, and it included the Mauj Collective. The show, however, almost did not come to pass, as the Mauj's inclusion was hotly contested by the directors of the Palace who, being used to presenting historical and traditional work, proclaimed their work was "not art" (Malik, interview). The Curator threatened to pull the entire show if the Mauj Collective was not included, and the directors relented, allowing the show to go forward. Their work is multi-genred and multimedia, incorporating calligraphy, sound, social and digital media, locative media, and video, as well as performance and VJ techniques. In a time and country where political art and personal activism can be deadly, art and activism are essential. The kind of knowledge the Mauj members teach and explore in their art opens the whole world to people in a time of restrictive censorship. It also creates a space for the discussion of ideas for which there are no other venues. Mahboob is the creator of a documentary called *The Colours of Sand* about environmental conditions in the desert of Thar and explores how environmental changes are affecting its inhabitants. Husain creates cell phone-based art and explores the nature of the diaspora in his philosophical writings and photography. Ahmed created a womb-like immersive virtual reality environment to help battered women explore and heal their pain. Another member of the collective, Amber Rana, is the creative director of Geo-TV, the national television station in Pakistan. When Asif Ali Zardari took control of Pakistan in the wake of Benazir Bhutto, his wife's, assassination, he imposed a 78-day media blackout. Rana and her staff were locked out (and without pay) during that time. Even though their own lives were in danger (and journalists 'disappeared' during this time), they waged a campaign to have the media reinstated by remixing the concept of posters for missing children. Creating printed flyers and posters, and websites, they ran a massive campaign featuring the faces of the station's news anchors listed as 'missing.' They did not cease until Zardari had reinstated the media two and a half months later, in large part due to their efforts. As of 2011, the Mauj Collective is largely disbanded, but the members have taken up solo careers and continue to work in new media. Malik believes that a lot of the work they did initially is now being replaced by mobile phone apps, and so now she and the other founding members can direct their energies in different directions to realize new initiatives and new projects (Malik, interview). The country is a militarized zone these days however, and as the walls get higher and higher, it gets harder and harder to find venues for art, digital or otherwise.

By contrast, in China political tensions are loosening up for artistic expression (or were until the Cultural Revolution-style return to Ai Weiwei's suppression that resulted in an 80-day jailing in the spring of 2011. So far all indications are that his imprisonment was an isolated incident regarding his outspoken politics rather than a new policy position toward artists). In 1978, Deng Xiaopeng's then new socio-economic policies started revitalizing China and unleashed massive urban growth and modernization, as well as reopening relations with the West. Advertising returned to China in 1979 (Wang, 2010, 1). The effects of the sweeping changes were slow to show at first, but by the 1990s they had transformed the nation entirely. One of the fallouts of Deng's globalization policies has been to resituate the Chinese art world as a player on the world stage. The first indicators of that were seen when three major international shows were staged in 1993: "Mao After Pop" in Australia, "China Avant-Garde: Countercurrents in Art and Culture" at the Venice Bienniale in the *Haus der Kulture der Welt* hosted by Berlin, and "China's New Art, Post 1989" on the island of Hong Kong (Hung, 2008, 404). That was just the tip of the tail of this sleeping dragon as China continues to gain in influence and power in the art world. Art critic Pi Li says that "this style" as revealed at the Venice Biennale, "which reflects non-Western ideology, established the label and standard for Chinese contemporary art, and also initiated the process of cultural understanding and exchange on that foundation" (Pi, 2009). It is now one of the hottest art markets on the planet, with individual pieces – usually paintings – selling for millions of dollars.

Due to violent political interventions – first the Communist Revolution in 1949, then the Cultural Revolution from 1966 to 1976, and finally the June Fourth Movement that culminated in the Tiananmen Square massacre in 1989 – Chinese art has reinvented itself from new ideological positions several times during the last century: "Each rupture has forced artists and intellectuals to reevaluate and reorient themselves. Instead of returning to a prior time and space, the projects they have developed after each rupture often testify to a different set of parameters and are governed by different temporality and spatiality" (Hung, 2008, 396). At roughly the same time as events in Tiananmen, an exhibition called *"Magiciens de la Terre"* opened in Paris at the Centre Pompidou, marking, Marianne Brouwer argues, "the arrival of globalism in contemporary art" (Brouwer, 2004, 3). The show proved very controversial (the Curator, Jean-Hubert Martin, lost his position later that year) as works had been gathered from all over the world but were presented uncontextualized (Brouwer, 2004, 3). While it introduced many international artists, including Chinese ones, to the West, it demonstrated the worst attributes of the global art market. That year (1989) also marked a split between the kind of work that artists were doing within China and what Chinese artists who lived outside of the country were creating, which was unofficial and unsanctioned. At home in Beijing, another show, *1989/China Avant-Garde* was shut down on its opening day when artists' performance pieces turned out to revolve around destroying their own work (Van Der Plas, 2004, 11). The "result of years of making underground and alternative art that deployed various media and disciplines":

Performance, video and conceptual art found their way to China at a time when Chinese society was gripped by a sense of hope and expectation. The Cultural Revolution was over, the borders had opened slightly and reality was becoming increasingly real. It was no longer a question of making standard work on Chinese art history. Instead artists set fire to their work (Xiamen, 1986) or threw the standard work on Chinese art history into a washing machine together with its Western equivalent *A Concise History of Modern Painting* and then publicly exhibited the crumpled results (Huang Yong Ping, 1987). This energetic cultural explosion reached both its zenith and its turning point at the *China Avant-Garde* show of 1989 (Van Der Plas, 2004, 11).

The political crackdown revealed by the Tiananmen Square massacre stopped most of these art trends and activities in their tracks, and sent art in China down a different road.

Art critic Wu Hung argues thereby that the modern art of the 1980s and the contemporary art of the 1990s were not merely sequential trends, but "disconnected endeavors conceived in separate temporal and spatial schemes" (Hung, 2011, 396). The artists of the first period aligned themselves with the European Enlightenment model put forward by the May Fourth Movement, which led to information overload or what Einstein would have called an 'information bomb' as they soaked up Western philosophy, critical theory, and modern art. The 1980s with its open-door policy marked the reintroduction of much that had been forbidden during the Cultural Revolution. Artists and theorists swallowed the whole modern art tradition in one piece resulting in a mashup of approaches and techniques as the full spectrum was gobbled down in a single bite: "The chronology and internal logic of this Western tradition became less important; what counted most was its diverse content as visual and intellectual stimuli for a hungry audience" (Hung, 2011, 397). Without a chronology or a logic of development these decontextualized movements were tried on for size and resituated in an Eastern context. According to Lü Peng and Yi Dan's *A History of Modern Art in China: 1979–89*, the most influential thinkers of the decade were Schopenhauer, Nietzsche, Sartre, Freud, Jung, Camus, and Eliot (Hung, 2011, 397). German curator Ulrike Münter says that Marcel Duchamp and Joseph Beuys were also very influential during this period in China and were frequently cited as sources of inspiration. Tiananmen changed all that.

In the wake of that cataclysmic event, the break with the preceding art tradition was nearly total:

These grand [Western] names suddenly become infinitely remote, and few Chinese artists, if any, continued to seek guidance from them. Rather, the sharp historical gap created by the Tiananmen incident distanced Chinese artists from the previous era, enabling them to develop a radically different relationship with history and with the surrounding world. In this process, they also disengaged themselves from *yundong*, the Chinese term for large-scale political, ideological, or artistic "campaigns" or "movements" (Hung, 2011, 397).

The *yundong* had been a powerful force which translated its influence into propaganda materials that celebrated the Cultural Revolution and sought to suppress expressions of individuality in life as much as in art. Even when

1980s art critiqued political ideologies of the regime, it had done so in mild tones from within the system.

The 1989 tragedy at Tiananmen was combined with another kind of revolution: economic reforms transformed Beijing and Shanghai, making them cosmopolitan gateways to the world, and with that change came an urban population with money to spend and (even more remarkably and transformatively) marked the arrival of privately owned art galleries. Riding the tide of these changes was what Hung calls the Chinese 'contemporary' art movement. Even in painting, the most traditional of art forms in China, political pop becomes a driving force and acquires a sense of humor: Zhang Hongtu can paint Chairman Quaker Oats ("Long Live Chairman Mao," Series #29), and Wang Guangyi, the father of political pop, can incorporate capitalist trademarks including Coca-Cola and Vuitton into the iconography of the earlier political fervent style. Shi Xinning can create alternate histories on canvas by inserting Mao into vintage celebrity culture news photos with Marilyn Monroe, into foreign spaces like the Oval Office with Elvis and Nixon ("White House), or have him gaze in fascination on Duchamp's urinal ("Duchamp Retrospective Exhibition," 2001).[6] The abandonment of the *yundong*, however, also led to the abandonment of Chinese artistic traditions right across the board. Western capitalism in turn seemed threatening, poised on the horizon to impose a new overarching ideology. Where painting and printmaking had been revered above all else, artists now began to work in multimedia, performance, and to create site-specific installations. The results of this were electrifying, Hung argues:

> new, experimental art forms provide contemporary Chinese artists with an 'international language.' Inside China, however, these forms have served to forge an independent field of art production, exhibition and criticism outside official and academic art. In denouncing painting, artists can effectively establish an 'outside' position for themselves, because what they reject is not just a particular art form or medium but an entire art system, including education, exhibition, publication, and employment (Hung, 2011, 399).

These changes in practice and attitude start to open the way for new philosophies and new media. The move away from a tradition-bound *yundong* led to an exploration of the parameters of self. *Yundong* had defined self-portraiture as "bourgeois" and "counterrevolutionary" (Hung, 2011, 404), but portraiture emerges in the 1990s as a new ironic and hyperbolic form as in Fan Lijun's paintings, Xu Jianping's urban installations of copied paper paintings of Van Gogh's self-portraits, and GI Joe and others (see Huot, 2000, 145–6), Zhang

Dali's graffiti, or Miao Xiaochun's virtual worlds. The self is shattered and adrift, both everyman and no man, solitary and alone. The notion of *the movement* becomes an ironic position.[7] Every artist becomes (potentially) his or her own movement and agent of artistic reform. There is no new visual iconography to replace the accepted and expected images of the model-driven Cultural Revolution and after, and so artists use their own image in their works as an indicator of an ideological absence until a new lexicon is defined. It is in these new forms of media that artists are most free to experiment both within and outside of the Chinese tradition. Gone are the days of faithfully painted replicas and pastiches of propaganda posters from the Cultural Revolution.

Political pop and cynical realism also brought with them new dangers that had long been apparent to Chinese artists: the ability of politics to manipulate their art. After the arrival of video art (*luxiang yishu*) in China in 1990, there is an awareness of

> the lure of commercial incentives and the peril of being manipulated by western cultural neocolonialism. Both dangers had their roots in old-fashioned mediums (sic) and methodologies. This was the background against which young Chinese artists of the time began to experiment with video art. They were searching for a new medium that could resist commercialism by western galleries, while at the same time provide a contrast to more official mainstream art. Video art was a medium that allowed for the expression of individual feelings and language, and it was also easy to use, disseminate and exchange. Under these circumstances, video art became their medium of choice (Pi, 2004, 15).

Digital technologies quickly became interwoven with video art, and it is often difficult to tell which is which just by looking. Narrative modes of storytelling also arrived with video art, and digital technologies introduced fantastic (i.e., nonrealistic) themes, special effects, and remixing, according to Pi. One of the earliest digital media artists, and one of China's most important conceptual artists is Wang Jianwei. He was the first Chinese to have his work shown at Documenta in 1997 (Whitaker). As a conceptual artist, his work is difficult to categorize as he switches between media and modes, including performance art, documentary videos, video-based installations, interventions, and social sculptures. He often explores what he calls "the third space" of the imagination, the Internet, or *Desert of the Real* (as one of his shows was called after Jean Baudrillard's theory) in his works. His digital video *Connection* (2000)[8] was composed of two video streams. One was a montage of erotic scenes from pirated American and non-Chinese Asian films – including the pottery wheel scene between Demi Moore and Patrick Swayze in *Ghost* (Zucker, 1990) and the scene with Leonardo DaCaprio and Kate Winslett on the prow of the Titanic (Cameron, 1997).[9] The stream incorporates not only erotica, but also folds in pirated foreign scenes of terrible violence and tremendous amounts of gore. Li says that out of this found footage Wang "spliced together" he "create[d] a 'new cinema'" (Pi, 2004, 16). On the facing wall, the second channel showed scenes of eight different Chinese families seeming to watch, across the gallery, the stolen images from films that would, in real life, be censored on their televisions in stunned disbelief. In the show "Synthetic Reality," Pi says that "the most significant aspect of his [Wang Jianwei's] work is his consistent use of convenient, low cost digital technology to create a 'new narrative' to challenge old narratives dominated by commerce and politics" (Pi, 2004,16). It is no accident that when Wang wanted to avoid the consumer culture model that the technologies availed, he made similar choices for lowbrow forms to shatter the spectacle as what his Western compadres did. In fact, the video quality in the A stream is so poor because it was drawn from the pirate VCDs families in the B stream actually watched. VCDs (video compact discs) were long the medium of choice in Asia because unlike DVDs they had read/write capabilities. The copies that circulated degraded as they were copied and recopied.

The first video art show in China was called "Phenomenon and Image" and it was held in 1996.[10] In "Chinese Contemporary Video Art," Pi says, "Video means 'reflection' and image indicates 'response'. Hence the reason why young artists chose video as the medium was that it is an art that embodies reflection" (2009, 304). The video artist who curated that exhibit, Qiu Zhijie, defined an equation for interactive art: "traditional media = reaction = virtual individual = perception = unilateral video = reflection = real individual = perception, imagination, action = interactive" (quoted in Pi,

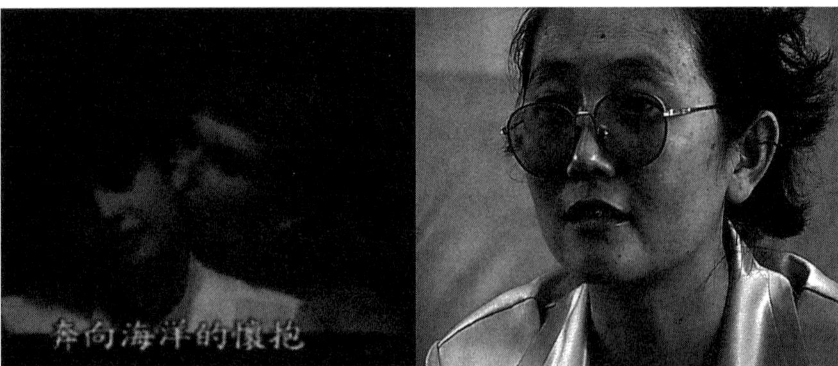

Figure 11 Wang Jianwei, *Connection* (2002)

2009, 305). Video art has different origins in the East and in the West. In the East it was an aesthetic choice based on the positively valued attributes of the medium; in the West it was a rebellion against traditional media (Pi, 2009, 305). The second video art show was held in 1997. The curator of that show, Wu Meichun, encouraged artists to develop their work. Three new schools or modes grew out of that show: the new documentary movement, a group making time-based and 3D narrative, and interactive video in which practitioners explore either the medium or the artist's interior world (Pi, 2009, 305–6). Artists quickly exhausted the possibilities for interactivity in video and turned to the more pliable properties of multimedia instead.

The first large-scale new media show in China, *Synthetic Reality*, was held in 2002. The work of the Generation of 85 (or the 85 Movement as it is sometimes called), it featured nine video and new media artists, and was all in Chinese. The language is significant because it meant it was a show designed only for home-grown consumption (although it did later travel to Amsterdam), but that it was designed to bring artists back "home" to show their work, according to Ni Haifeng (2002). The artists organized the show themselves without a curator, and drew heavily upon historical materials that would have particular resonance for those who had lived through the Cultural Revolution and the Tiananmen tragedy. The artists who participated were among the best and brightest working in new media: Geng Jianyi, Zhu Jia, Shi Yong, Wang Gongxin, Chen Shaoxiong, Long Yongbin, and Wang Jianwei, as well as two artists whose work I discussed earlier, Zhang Peili and Ni Haifeng. Ni and Geng documented the event. Ni included work and wrote the website copy, and Geng's contribution was documentation-as-medium; that is, documentary works about each artist and their reason for participating in the show.

Russian painter Anton Skorubsky Kandinsky dubs this rabid creative outpouring in both new and more traditional media "China-ism". It is the product of an incongruous crash of Western and Eastern materials combined with ill-fitting translations. Historically, in the early days of Communism, Chinese art and culture developed under the tutelage of Russian artists and the Western European school. In a smart and ironic reversal, this famous grandson of Wassily Kandinsky finds tremendous inspiration in the contemporary Chinese artist community's revisitations of pop art. Believing that Warhol had introduced the West to "China-ism," Kaninsky says:

Contemporary Russian art is very simple. It copies Western contemporary art, sometimes producing very weak results. At the same time, contemporary Chinese art has made it in the West as a completely independent and authentic event. I don't understand how Mao can simultaneously be in a mausoleum and on the paintings and still be received normally by the public. Contemporary Chinese art has very strong visual hieroglyph-symbols. In the West, the strongest visual symbols are the Mona Lisa, the Statue of Liberty, the Sphinx, and Michelangelo's statue of David. Meanwhile in China, we see Yu Minjun's smile. We think that China is primitive and communist but in reality they laugh at us. Yu Minjun's smile is a very strong symbol! (Kandinsky, 2010).

Dismissing the commercialism of contemporary Western artists like Damien Hirst, Kandinsky says, by contrast:

Do not think that China-ism is something unique or applicable only to China. China-ism is important for all of us, non-Chinese people especially, because contemporary Chinese art can contribute to the development of art internationally. This is a historical moment! China-ism is a movement for non-Asian artists that will allow them to contribute to the advancement of world culture and art (Kandinsky, 2010).

Kandinsky's own witty paintings use Chinese artists like Shi as inspiration, but there all resemblance ends. Kandinsky situates Warhol with Mao as an adoring subject ("I Want To Be Mao Zedong") and positions Chuck Close next to Ai Weiwei – depicted as President of China – in "I Don't Want to Be An American Artist. I Want To Be A Chinese Artist."

Ni Haifeng, the artist who created the reproductions of hand-made Dutch china in *Of the Departure and the Arrival* discussed in the Introduction, sees his own production process as being rooted in a kind of "cultural translation" as well. In that installation work of counterfeited crockery, Ni says,

the Western objects with their own history and cultural significance are not only reproduced, but translated and negotiated culturally by the laborers in China. The resulting objects are neither Western nor Chinese; they are a transposed hybrid. Sometimes it's a more complex multi-directional translation. For example, in Shrinkage 10%, the original Chinese porcelain objects from a Western museum are re-translated by Chinese workers into a set of diminishing copies. I use the word 're-translate', because the originally Chinese objects were 'museologically translated' from colonial trophies into iconographic Western museum objects. The re-translation takes a reversed route, from Western museum objects back into, with a twist, contemporary Chinese counterfeits (Ni Haifeng and Yao, 2011).

What quickly becomes clear upon a close examination of this new generation's adoption, adaptation, and appropriation of Western icons, images, and ideas is that there is no one-to-one correspondence, straight lines, or easily mappable trajectories. Instead these hybrid spaces are complex, fertile cross-pollinations, cultural resituations, and willful misreadings. Where the startlingly alien nature of these new works and stark nature of their originality becomes really apparent is in digital media.

Digital media has been a cautious arrival on the art scene in China and it still has only a handful of prominent practitioners in the realm of digital art. Of these Miao Xiaochun, Feng Mengbo, and Cao Fei stand out as inspired, creative originals. Miao is a photographer who crafts virtual, 3D worlds from the landscapes of traditional paintings that he then populates with contemporary imagery and models of himself. Feng Mengbo has reimagined the video game as a space for the exploration of self, subjectivity, surveillance, nation, and combat. Cao has crafted an entire virtual city, which is populated with hip-hop artists and cosplay characters. She samples the real world to craft a virtual one. Miao's decontextualized images are in some ways more akin to readymades rather than to remixes. In fact Chinese adaptation in contemporary art in general "places a strong emphasis on its aspects of decontextualisation, the appearance of something in a foreign or alien environment, and the defining role of the spectator for the work of art," according to German art critic Ulrike Münter. As a photographer Miao is always aware of perspective and multiple and shifting points of view. Choosing European masters and their masterpieces as source material, Miao transforms the medieval or renaissance vision within entirely new 3D computer-generated worlds that he uses both as a subject for photography (produced as C-prints from multiple angles) and for computer animation. He was not the first photographer to use this technique (this is the source of the art form known as machinima, for instance, or see, for example, Toronto artist Ray Caesar's work), but he does it exceptionally well. Miao burst on the digital media scene in 2005 with the first of his trilogy of European masters series, using Michelangelo, Lucas Cranach (the Elder), and Hieronymous Bosch as his departure points in exploring possible futures. His first work of this sort was the remix of the famous fresco, Michelangelo's *Last Judgment* (1536–41). Miao's *Last Judgment in Cyberspace* (2005) re-created the world with its 400+ figures as a virtual environment with a naked scan of himself standing in for all of the figures. Michelangelo's painting was harshly criticized in its day for its nudity (the draperies were added by another painter in Michelangelo's lifetime), which fulfilled a leveling function. Nudity put everyone on the same level and introduced equality into saints' and sinners' chances of entering into heaven. His "stunning thematically, technically, and artistically complex work"

(Panhans-Bühler) is decontextualized in every way with a naked, gray-scale Chinese male standing in for all genders and beings:

> Miao Xiaochun decided on one single figure as their basis: his own person, stripped of socio-cultural and temporal characteristics, that is, stripped of clothing. In doing so, he placed in the forefront a nakedness – especially as a Chinese – that even Michelangelo's provocative nudity remains in the shadows, all the while avoiding any idealized reverence for the body itself (Panhans-Bühler, 2010).

Christ, Minos, Charon – they all become Miao, rending holy spirit, angel, saint, and devil in the guise of a single figure. By erasing gender and stripping one of the most power symbols of Euro-Christian worldview of its theology and its iconography, and by altering the race of the figures, Miao's depoliticized selves become instead a statement on Communist ideology (post-Mao) and the trademark traditional 'selfless' perspective of bygone Confucian ideals. With a cameraman's eye, Miao's *Last Judgment* re-creates Michelangelo's vision in the round. Figures become three-dimensional and the world is generated for us as viewers not just through an animated fly-by video, but also as photographs taken from three perspectives: horizontal, bottom, and side. In the horizontal view, the Christian vertical iconography is spun sideways. Zielinski says that this is a metaphor for information overload (Zielinski, 2010). Instead we get a photographic panoramic view of heaven with a single, prominent image of Miao in the right foreground as lone creator, taking a virtual 'picture' with his hands. Where a Western artist might use his own image as objective material for the creation of a work, Chinese artists like Miao use their own image as the subject of their art. Using the process of taking screen shots as in machinima (a form of cinema made in game spaces and Second Life), the viewer is inserted into the work. Miao, however, chooses strange angles and so we see this world upside down as devils would or from other perspectives from which it was never meant to be seen. Huang Du (Curator of *Microcosm*) says that the strange perspectives in Miao's paintings actually refer back to traditional Chinese storytelling where a holistic sense of the world was conveyed by a character appearing in different places in the story. Miao says that his perspectives also refer to Asian notions of perspective, which were so different from Western ones. Traditional Chinese perspective lets one see a scene from multiple angles: a bridge might be depicted in such a way that both top and bottom are visible at the same time. In the same way, Miao says that characters were constructed in paintings in the same way: "In ancient Chinese paintings, the same figure can appear in different places, such as walking from bridge to pavilion, drinking with friends there and then climbing up a mountain… people's behaviors at different times can all be put on a scroll" (quoted in Huang Du, 2010).

In 2007, Miao unveiled his next work, "H20: Genesis," a work concerned with works that create or are concerned with watery worlds. This work is also composed of C-prints and an animation again. The animation is a mashup of multiple paintings: Lucas Cranach the Elder's *The Fountain of Youth* (1546), Michelangelo Buonarroti's *The Deluge* (1508–9), which explore birth and destruction; next come Tiziano Vecellio Titian's *Bacchanal* (c.1518) representing desire and release, and Giotto di Bondone's *Scenes from the Life of Christ: Washing of Feet* (1304–6), Antonio and Piero del Pollaiuolo's *The Martyrdom of St Sebastian* (c.1475) (with the martyr spouting water from his wounds instead of arrows), where some figures have been transformed into vessels of water; in Pieter Bruegel the Elder's *Christ Carrying the Cross* (1564), where Christ falters under the cross, and transforms the holiest figures – Mary and other saints – into shapes of pure water; the last group repurposes Nicolas Poussin's *Landscape with Diogenes* (c.1647), and depicts the philosopher Diogenes who throws away his last possession – a cup – when he sees a man drinking from a stream using his hands, and ends with Michelangelo's *The Creation of Adam* (1510).

The starting point for the work, according to Ursula Panhans-Bühler, was Miao's "fascination with Lucas Cranach the Elder's painting *Fountain of Youth*, an iconographically singular work from the middle of the 16th century that transforms the Christian, religious idea of the fountain of youth into a heathen classical antiquity" (Panhans-Bühler, 2010). A painting about life, death, metamorphosis, and rejuvenation dating from 1546, Cranach's original is a triptych of old age, rebirth, and beauty. It is also possibly the strangest choice for one of Miao's remixes. Old women populate the left-hand side of the painting from where they enter the healing waters. In the fountain in the center, they are transformed, only to exit on the right reborn as ideal young, female beauties. An iconic image about transformation and the fleeting nature

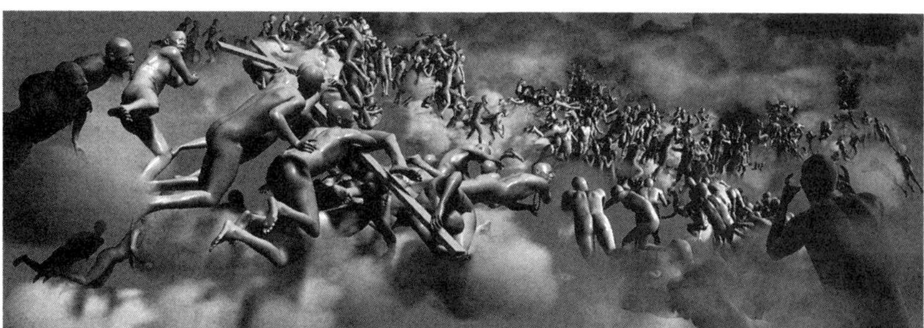

Figure 12 Miao Xiaochun, *The Last Judgment in Cyberspace*

of youth, Miao transforms it utterly. Where women, critics, and scholars of the Twentieth and Twenty-first centuries have fought so hard to reinsert women back into art history as more than models or objects, here she is not even merely neutered, but erased completely. Gender and racial politics are, of course, underlined and made all the more prominent for being rendered in the singular body of Miao, as stand-in for the type, Asian male. Curator Siegried Zielinski says,

> By taking his own exact measurements and inserting a calculated image of his physiognomy in place of [a] painting – a kind of geometrical physiogram – Miao Xiaochun has taken the dissolution of the Renaissance ideal to its radical extreme. In the world of clones, there are no longer any originals. Behind the masks of sameness there remains only a suspicion of something consistent that, however, can no longer be identical with itself. *Always the same, but never myself,* (Zielinksi, 2010).

To transform a painting that is so deeply concerned with women's reproductive roles and the politics of gender, and to resituate it in an all-male landscape changes the meaning of the image absolutely, especially since Miao's figure never ages, but is always identical in this world. Becoming a metaphor for the copy itself in an age of perfect replicas, Miao's cloned image is about the perfect reproducibility of data. It is also a cautionary tale, becoming a statement on sterile ecosystems, and the dangers of re-creating the world in our own image. The subject is disposable, easily replicated in a digital age – all and nothing. The creator of worlds, Miao wears the eyes of Venus as well at the center of the image, still holding her child Cupid aloft. The transformation in the work is in the movement from one painting to another, in the history of painting and styles and works. Maio's work threatens to become a statement not on transformation, but on stasis and a dehumanized universe. The endless repetition stops us, and reminds us how incomplete a desexualized male-only vision is and how essential the female element is in the ecology of the world. Is this a formula for repetition or a creative remix? Miao says in his artist's statement that he owns art history and that he rewrites it in his own image:

> [U]nlike *The Last Judgment in Cyberspace*, the meaning of this image has changed from unifying individual characters in a single work to connecting various historical paintings into "a kind of *metabolism*" (Miao Xiaochun's words). In his view, if water flows from one organism to another in the natural world, then this project makes him (or his image) a neutral element "flowing" through works of art created in different times—a process which generates new works based on old themes (Wu Hung, 2011).

Miao-as-copy has become the subject of the work itself. Zielinski (2010) says of *Last Judgment*:

> The time after the subject has recognized that it was never complete and assumes that it never will be whole. The only originary aspect of Miao Xiaochun's *Last Judgment* is the possible relationship of the individual cloned characters to one another, a relation that the artist can randomly alter. The particular, the sensational has slipped to the relational, become flexible, black and white, an abstraction. From here, after the *Last Judgment* the artist will push forward to the equally colorful *Garden of Delights*.

The *Garden of Earthly Delights* is purposefully mistranslated into *Microcosm* by Miao. It is a 14-minute animation with C-Prints and a nine-panel screen. It reimagines Hieronymous Bosch's painting, Leonardo da Vinci's *The Last Supper,* and some of Naim June Paik's work, with all characters in the image of Miao. In the animation, some of the characters change places, but are not animate. The soundtrack is the sound of a keyboard and the ambient sounds of destruction: breaking glass, rolling tanks, and what Zielinski calls "a powerful Wagnerian soundtrack." The animated world is a remix. Not understanding the Christian iconography (and not caring to), Miao simply deletes the objects in Bosch's world and replaces them with contemporary items: stop signs and a laptop (with Miao's face) and helicopters and airplanes from stock photography. Miao describes his mistranslation of Bosch as a representation of "the dreams of modern people" (quoted in Huang Du, 2010). He explores contemporary issues like the environmental crises, war, famine, and death in this complex work. In *The Dance of Death,* for instance, he feels that he needs to reflect humanity's ability to wipe itself out with the press of a button on a computer, and so the figure of death is replaced by a laptop with Miao's face, "Death dancing, nuclear war starting, total destruction" (quoted in Huang Du, 2010).

Beijing artist Feng Mengbo was catapulted to international fame in 1994 when he produced his own screenshots in a show called *Game Over: Long March.* He did not own a computer, but he was an avid arcade game player. In *Game Over: Long March,* he remixed a world that might exist if there were Chinese video games – games that used the traditional Communist iconography of painting. In that show, Feng's 40 canvases of signature moments in Chinese contemporary history, images of Mao, and game images were all painted as if they had been created in eight-bit animation and as if they were some distant Asian cousin of Super Mario. A companion show, "Taxi! Taxi! Mao Zedong," translated Mao, making his customary salute to the troops, into the act of hailing a cab. Again inserted in game space on canvas, Mao hails cabs in the eight-bit game spaces of Beijing, but never catches one. Those paintings won Feng a space at the Venice Biennale. While in Venice, he bought his first computer and it was then that he could start creating the digital environments and game mashups he had imagined. His first interactive work was a hypertext called *Private Album* (1996), which told his personal history in multimedia; he folded in found footage (documentary, historical and otherwise), images of books, and Beijing opera. *Mount Doom + Taking Tiger Mountain by Strategy* (1999) followed; it was a three-channel animation. One stream showed 42 film clips from the Chinese film *Taking Tiger Mountain*. The second projection showed a very fast – one-minute – screen capture of the game doom. The third projection was images of books. His next three works all revolved around the computer game *Quake*: *Q3*, a machinima movie make in *Quake*, *Q4U*, and *RQ*. *Q4U* is remarkable as an interactive video game that superimposed Feng's face on the first-person shooter's persona and on his opponent's person. He carries both camera and gun, and when the user 'wins' they obliterate him. His choice of using his own image revolved around Chinese iconology. In a post-Tiananmen world, artists can no longer use the image of Mao as a stand-in for the individual and everyman in the way that they used to. In a capitalist China, no one knows who they should be or what face they should put forward. Feng used his own image as the most ethical choice so that he would not have to pay royalties to someone else, and so that any violence that is done is only done to him.

Feng is most celebrated for his massive 80-ft x 20-ft work *Long March: Restart* (2008; see book cover). Created in multimedia fusion and programmed from scratch, *Long March: Restart* plays and feels like *Super Mario Brothers* or *Donkey Kong*. The cultural framework, however, is entirely different. The work moves through 14 levels of key times in Chinese post-Communist Revolution history – from the Forbidden Palace to being a Chinese alien-fighting astronaut in space. All of the characters have been translated into culturally appropriate clothing. The game is fully functional and can take hours for a user to play in gallery space. This takes gaming to a new level of cultural translation, reinventing the game through mistranslation and international interactivity, for the work has been purchased by the Museum of Modern Art. You too can be a Chinese soldier in gallery space in New York.

Another game interface that has been exhibited in China (at Trans-art Week at the Xi'an Academy of Fine Art, July 2–16, 2010) and that inserts the user into the work in dramatic ways is Canadian-U.K. artist Steve Gibson's game mod *Grand Theft Bicycle*, a game installation (2007). In sharp contrast to the Second Front's performance of *Grand Theft Avatar* and to Feng's ground-up re-engineering of a game interface for his own ends, Gibson (who is on the

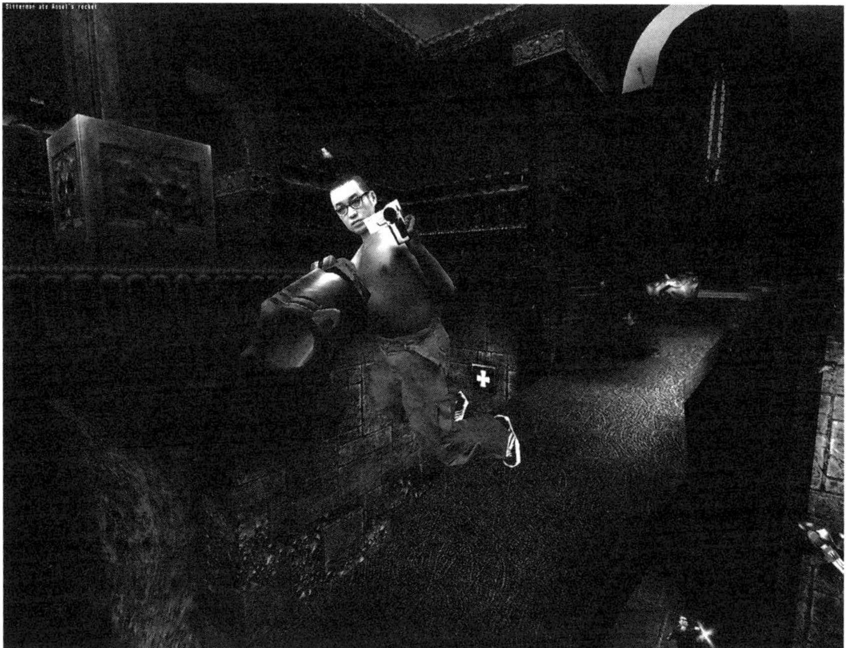

Figure 13 Feng Mengbo, *Q4U*

faculty of the Central Academy of Fine Art) actually hacked a *Grand Theft Auto: San Andreas* game cube using a game engine. The closed source original game had many copyright and hacking protections in place and so it was a complex task to enable the game to become a foundational text. Programmer Justin Love manipulated the source code that extracted the original characters and replaced them with prominent politicians and terrorist leaders, including George W. Bush, Margaret Thatcher, Fidel Castro, Yassir Arafat, Kim Jong-Li, and Arnold Schwarzenegger. He also manipulated San Andreas so that it appears to be modern-day Baghdad. An interactive and immersive installation, the analog army-drab green bicycle acts like a joystick to the digital interface, which is projected on a large screen (a minimum of 4 x 3 metres) in gallery space with a two-speaker immersive sound environment. The 'Borgcycle' is a 1950s-style bicycle, sensor-modified, mounted in a stand in the back and a spinning circular platform at the front. It may be used with any joystick-driven PC game.

The bicycle is equipped with sensors, control buttons and an analog-to-digital interface that essentially transform it into a joystick. The Borgcycle faces a large projection screen and is set up so that users can "pedal into" a 3D gaming environment, a recreation of a desert city (Baghdad). The characters of the original version of *Grand Theft Auto* have been modded into familiar political figures who form warring gangs. The invaders include George Bush, George W. Bush, Dick Cheney, Condoleezza Rice, Margaret Thatcher and Tony Blair. The insurgents include Saddam Hussein, Yassar Arafat, Osama bin Laden and Kim Jong-il.

The player actually rides (or appears to ride) into the screen-defined game space. By transposing the original game space to a virtual version of the real political world complicates the fact that the player has to choose whether she wants to be an invader, an insurgent, or an onlooker. Gags are also built into the game, like it has its own product placement opportunities, promoting "Freedom Cola." "*Grand Theft Bicycle* is meant as an ironic commentary on the connection between gaming and war, but it remains an ass-kicking shooter game that will let you vent your anger at the idiots who are in command" (*GTB* Documentation). The further the player progresses through the game, the more deadly the bicycle's weapons become. The final level grants the player weapons of mass destruction. Particular players will defend other players; for instance, "Tony Blair'll take one for George W.; Saddam Hussein will drop down in front of Osama" (*GTB* Documentation).

The celebrated young artist Cao Fei performs a different kind of play and translation of the real in her profoundly original works. Translating and adapting the role-play practices of the cosplay world, she took her personal avatar and Second Life world RMB City all the way to the 52nd Venice Biennale. The world is so fascinating because of the way she blends East and West, Chinese philosophy and Western capitalism, youth culture and

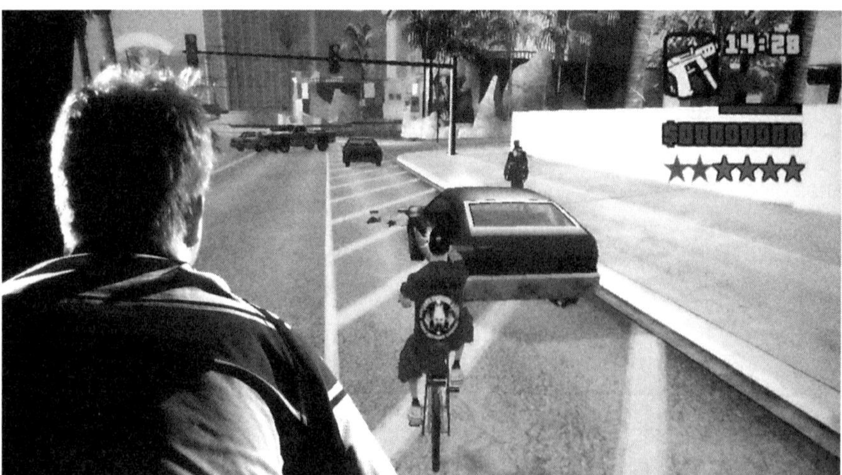

Figure 14 Steve Gibson, *Grand Theft Bicycle*

a sober analysis of the nature of both real and virtual worlds. It stands "as mirror between the rl and the SL, reflecting the world and China under the fourth wave" (Cao, 2011). Part of her display in Venice was her 28-minute Second Life machinima documentary *iMirror* (2007), which was filmed on her virtual island City. The work transcends many barriers and states, existing in Second Life and in real life in a giant inflatable multi-domed igloo-like pavilion at the Biennale. To view the documentary in the Pavilion, the audience had to climb into an elevated cup similar to the one seen in the film. *iMirror* is an astounding piece of work. The music is by Prague and the lyrics by Octavio Paz (in translation). Filmed in three parts, Part 1 is about world creation and explores China Tracy's (Cao's avatar) new world and the culture, and the translation or mistranslation of China into this Western space. Part 2 is the story of her avatar China's encounter and friendship with Hug Yue (Ed Mead), a 64-year-old man from San Francisco who understands what she is feeling. Part 3 is an exploration of identity formation in youth culture, countercultural subjects, avatar building, entertainment, and capitalist structures. The film is always concerned with finding a way to blend real and virtual lives in a way that makes sense for the user.

Part 2 seems to be building a romance, but Hug's real identity as an old man is gradually revealed when he puts on his old man avatar. The experience is underlined by China as a virtual one. As she was aware from the beginning, this was a relationship where "nothing happened, nothing will happen." The world is not static; far from it, but it is not real. The real part of Second Life is 'us.' We bring ourselves into the fantastic space and our feelings fill it. The middle part of the film reveals that China is all too aware of reality. "Some eyes watching us," she tells Hug. "We all in the film." They discuss the merging of the two worlds and how hard it is to separate their feelings in these parallel dimensions. Second Life is a kind of drug that lets them forget their pain for a while. In Second Life (unlike the real world), each of them does not have to act, but can just be who they are. Their avatars are their mirror reflecting their 'real' selves back to them. When China asks Hug what he thinks about the digital world, he says that "It is one that is dominated by youth, beauty and money. And it is all an illusion." Like his own identity.

Part 3 of *iMirror*, as an exploration of capitalism and identity, is the most interesting part of all. Looking for all the world like Andy Warhol's *The 13 Most Beautiful Boys* or the Mattes' *13 Most Beautiful Avatars,* Cao explores simultaneously the glamor and the emptiness that the multitude – this creative youth culture in our time – feel. Hip-hop characters, punks, fashionistas, animals, Goths, cyborgs, cosplay characters, all populate this limbo and all are equally without direction. Capturing perfectly the malaise of a mediated generation, Cao demonstrates how they do not separate fantasy from reality, but that they are far too cynical to buy into it wholly either. As they craft themselves into exquisitely beautiful images, take virtual lovers, dance virtual dances, hold virtual weddings, gamble with virtual money, and smoke virtual cigarettes, China is always aware of how hollow these lonely souls are. In Second Life, "We are not what we originally are and yet we remain unchanged," she says. "Hopefully, there's new possibility of combination in an electronic life, a new force which transcend this mortal coil." China, a tourist just like us, is our narrator, anti-heroine, master, and guide. She tells us, "To go virtual is the only way to forget about the real darkness."

Perhaps more than any other artist I have examined in these pages, Cao gets to the heart of how the real and virtual worlds – how real world and digital cultures – are mingling. Solid lines can no longer be drawn between creative space and the real world, between play and work, between creative practice and criminal acts of piracy. What is imagination if not the creation of utopian or dystopian spaces where we can try on alternative practices, selves, and worlds for size? Posing new possibilities for real and virtual relationships and interfaces with the world, Cao blends past, future, dream, and dialog into a fruitful and productive creative space for the development of a political consciousness. And develop it she did. It is hard to see the seeds of her later politically hard-hitting piece *Shadow Life* in these dreamy spaces. As the virtual world withers away, revolution marches both in capitalism's wake and against capitalist suppression of creative practice across the globe. Carrying China into a globally energized space, Cao creates a space where it might be possible to design new worlds outside of the constrictive parameters of the multinational-serving nation state. Her RMB City remains a creative space for the realization of new work. For instance, the project *People's Limbo* was inaugurated on May 22, 2009 in the Hong Kong Museum of Art as part of the *Louis Vuitton: Passion for Creation* group exhibition. In it, users debate the nature of capitalism under digital culture. (The ironies of this are extreme and provocative, of course, given Louis Vuitton's history as a high-profile manufacturer of 'fake' versions of its own handbags and other products.) In *People's Limbo,* one avatar named Lennon tells Mao that "this is not the era of revolution, nor is it an era of sex, but an era of capital." Marx responds that "Capital has allowed love to become a transaction." In virtual space there is no feeling, but, Mao remarks, "perpetual exercise preserves youth forever." What Cao reveals to us in her provocative works (works that are interesting not because they are in Second Life but because they are about the gap between the real and the virtual) is that there is life after art-as-documentation, but that life-as-documentation poses new creative possibilities for the disaffected multitude to critically engage with the world at a safe distance in RMB City's squares until such time as they are ready for full-scale embodied artivism to take back the square – be it Tiananmen, Tahrir, Trafalgar, or Liberty.

Conclusion

In the catalog of the first digital media art show in China in 2002, respected curator Pi Li wrote,

> Video art has a duty to critique our information culture, but this is not by any means its only duty. Contemporary art, with its emphasis on individuality and craftsmanship, is by its very nature ill-suited to hold any real dialogue with the commercialized mass media, much less act as its driving force. The relationship between mass media and video art, particularly as it was practiced prior to 1968, is similar to that of a flyswatter and a particularly stubborn fly. Now, Chinese artists, like their international contemporaries, have begun to realize that the only place they will find their own territory and style is in a place well beyond the reach of the authority of the mass media (Pi, 2009, 17).

More and more, by stepping outside of the venues of commercial culture and mass media, artists and activists around the globe are seizing the materials at hand and repurposing their tools to their own ends. The art that they make or the movements that they start are designed to foster dialog. Mass media culture belongs to no one and is all too easily coopted by powerful forces to be deemed trustworthy any more. The future of culture depends on people translating the materials in their midst into materials that enable them to tell their own stories and to share embodied perspectives. Science fiction author Neil Gaiman says, "We have the right, and the obligation, to tell old stories in our own ways, because they are our stories" (quoted in Jenkins, 2009, 109). Digital media, social media, and the powerful digital tools we have at our fingertips enable us to do just that. Digital prohibition is upon us and, just like those crazy speakeasy times in the early days of the twentieth century, people's creative expression will not be denied and, for a time at least, we have some of the most powerful distribution tools ever invented at hand and free to use. Activism does not happen in digital spaces, but digital technologies and social media enable people to gather and express themselves, and to share their art. Dutch artist and critic Florian Cramer says in his article "The Fiction of the Creative Industries" (2011),

For young people, TV has been killed by YouTube, the music industry by mp3, DVD profits by bittorrent, newspapers by the web. But even more significant than these shifts of consumer technology was the digital revolution of production. Most musicians no longer need a record label, but can master their music on a laptop. Thanks to the last generation of inexpensive digital cameras, cinematic films can now be shot and edited at home by freelancers. Writers no longer need publishers, but often are better off self-publishing via print-on-demand and e-books. In all these areas, "creatives" become allrounders. Division of labor is decreasing, not increasing, with many industries, big agencies and highly staffed bureaus becoming dinosaurs of the past.

In those dusty bureaus a whole new mode of creative expression has quickened into life and will not be silenced. Creativity does not care what medium it is expressed in, but the availability of digital tools is fostering a creative renaissance – despite draconian copyright laws – that will not stop. Britain has taken the lead in this ongoing Prohibition by striking out against absurd over-regulation and crippling fines. As the effects of legal ripping for personal use start to be felt, England, like China, may start to take a newly re-energized role in fostering new creative digital works – as we have already seen from Christine Wilks and Christian Marclay. By not enforcing piracy laws on the books in China and by defining digital creative practice as a legal right in the U.K., these countries are recognizing that to remix is human and to re-create divine. Let us hope that the rest of the world comes to its senses soon too.

Notes

Introduction: Ambivalence and authorship

1 Hardt and Negri resist defining the multitude on its own, but prefer to map it as a series of contingencies on the basis of its collective actions. Deemed by them to be a positive force within democratic contexts, the group is immanent, revolutionary and unmediated. In *Empire*, they map the potentialities for a new global constitution as "[n]ew figures of struggle and new subjectivities" surface from the multitude itself (2000, 61). The militant acts that we are seeing, they say, are all decisive signs of the politicized multitude at work, in this case an over-educated and underemployed one. "[T]he action of the multitude becomes political primarily when it begins to confront directly and with an adequate consciousness of the central repressive operations of Empire" (2000, 399). This is a productive force that sustains Empire even as it seeks its destruction (2000, 411). Hardt and Negri predict that there will be a catalyst that will ignite the multitude to self-organize and to seize – to reappropriate – wealth from the ruling class:

> This is when the political is really affirmed – when the genesis is complete and self-valorization, the cooperative convergence of subjects, and the proletarian management of production becomes a constituent power.[…] [They say,] We do not have any models to offer for this event. Only the multitude through its practical experimentation will offer the models and determine when and how the possible becomes real (2000, 411).

In their sequel, *Multitude: War and Democracy in the Age of Empire*, they see the multitude as a strategy (or a movement – or a strategic movement) of postmodern warfare with a genealogy derived from contemporary resistance groups.

2 A major innovation in the #occupywallstreet protest was the creation of a new social media app called Vibe. It resembles Twitter but is anonymous, temporal and location-based. The app, created by Hazem Sayed, is revolutionary because it gives users the ability to set a time frame (as little as 15 minutes and as long as 30 days) and a locative parameter (as little as 160 feet to global) for message distribution. As a tool for protesters, it gives them a more fluid ability to dodge police who are monitoring social media as events are happening.

The third space of authorship

1 It is not without some irony that it was selected in 2010 as a finalist in the first YouTube Biennale at the Guggenheim, called YouTube Play. I will discuss this work in that context, and the show as a whole later in the book.

2 Elayne Zalis. Reported on Facebook (July 1, 2010).

The new Prohibition

1 I have heard several different versions of this story from my family. I do not know if any of them are actually true.

2 By contrast, in the U.K., the tide has started to turn in the other direction. A ruling in August 2011 deemed that "copyright currently over-regulates to the detriment of the U.K." and it made ripping for personal use legal (Lee, 2011). More importantly, it advocated for the personal use of copyrighted materials as a *legal right* to which every citizen was entitled. The ruling said that "the widest possible exceptions to copyright within the existing EU framework are likely to be beneficial to the U.K." (Lee, 2011).

3 This is similar to the Ritilin problem in North America. If the vast majority of boys under a certain age are on Ritilin, then what attributes of the male gender are they trying to medicate away?

4 The Steve Kurtz case and subsequent trial is a case in point. Founder of the Critical Art Ensemble theater troop, Kurtz had been a long-time critic of social issues around bio-technologies. When his wife died in her sleep, emergency services came to his home and saw his bioart. The FBI was called in and Kurtz was imprisoned and interrogated for the better part of a day. All of his technology, papers, and books were seized, and his cat and wife's body impounded. Although reputedly suspected of bio-terrorism, the grand jury refused to bring these charges against him. Eventually all charges against him were dropped, having been ruled "insufficient in its face": a ruling meaning that even if the charges were true, no crime had been committed. See Lynn Hershman Leeson's film *Strange Culture* (2007) for an in-depth exploration of the case.

5 Pharmaceutical companies are infamous for their greed and for wanting it both ways. Copyright and patent law have allowed them to 'legally' cannibalize traditional cultures, substances, plants, and medicines in Third World countries and reap the profits from those (at best) borrowings, and (at worst) thefts.

Part One: The aesthetics of appropriation

1 Guy Debord spoke of strategies, McLuhan of probes, and tactical media artists used tactics. Agamben's atactics may pose an alternative for the age of the remix that requires methods for creating disturbances after rebellion and revolution have been proved to be impossible.

2 Miriam C. Cooper's *King Kong* (1933) is in an entirely different category, as it never leaves anyone in doubt that it sympathizes with the beast.

3 Lest you believe that movies are the only things that have revisitations, remakes, and sequels, think Volkswagen Beetle or the Austin Mini.

4 For an in-depth discussion of video's implications, see Kaizen, 2008, 257–72.

5 The collective was formed in 2004 and comprises Paololuca Barbieri Marchi, Alberto Caffarelli, Matteo Erenbourg, Andrea Masu, and Giacomo Porfiri.

6 Caronia is referring to Vilém Flusser's concept of "new 'original' and 'quiet images'" that have been transformed on both the spatial and temporal dimensions (Caronia, 2010, n.p.).

7 This was also the case until recently with mainstream publishers as well, where a select few authors are very successful and handsomely rewarded while most receive a pittance. Publishers are now on the decline (despite increasing book sales) and even Amazon has set up a self-publishing program for the Kindle, its electronic book reader, called CreateSpace.

Interruption (stoppage + repetition)

1 What is the role of the spectator versus a creator? Of a spectacle versus a work of art? How does this change the idea of the artist? Sontag seems to suggest that the remix is necessary in *On Photography* (1977).

> A way of certifying experience, taking photographs is also a way of refusing it – by limiting experience to a search for the photogenic, by converting experience into an image, a souvenir. Travel becomes a strategy for accumulating photographs.... Most tourists feel compelled to put the camera between themselves and whatever is remarkable that they encounter. Unsure of other responses, they take a picture (quoted in Ritchin, 2009, 53)

2 This too is the purpose of the freeze frame in film, which interrogates the nature of time and space in the moving image.

3 There are also innovative uses of the long take in new media designed to trouble the spectacular nature of film. A telling example is Canadian video installation artist Kevin Schmidt's *Epic Journey* (2010), which is a three-take 12-hour mashup of Peter Jackson's epic *Lord of the Rings* trilogy and the artist's long takes of a ride in a small aluminum boat down the Fraser River and through the rising industrialization of the lower mainland of British Columbia.

4 For an in-depth discussion, see Alexander Alberro's "The Gap Between Film and Installation Art" in Tanya Leighton's *Art and the Moving Image* (Alberro, 2008, 423–9).

5 Dubbed "Germany's best unknown filmmaker" by Thomas Elsaesser (2002).

6 Cited, for instance, by Siva Vaidhyanathan in the documentary film *Good Copy, Bad Copy* (Directors Andreas Johnsen, Ralf Christensen, and Henrik Moltke, Denmark, 2007). Online: http://www.goodcopybadcopy.net.

7 Experimental digital media can work with these techniques and new film footage to markedly different effect. Compare, for instance, "Granular Synthesis: Pole" by Ulf Langheinrich and Kurt Hentschlager, which juxtaposes multiple screens of the same video image in a state of perpetual interruption. The stuttering of the visual image is the result of a MIDI-based sonic manipulation they call "audiovisual re-syntheticizing." Through the granulation and gerrymandering of fragments of sound and image, they create "visual and acoustic symphonies" (http://www.mediankunstnetz.de/artist/granularsynthesis/biography).

8 As I will discuss at more length in further chapters, in particular, in reference to Theodor Adorno and Emile Horkheimer's "The Culture Industry: Enlightenment as Mass Deception" (1944).

9 In a communist context with a new capitalist economy, Cao Fei has created a multi-media theatrical work called *PRD Anti-Heroes.* It remixes Hong Kong melodrama, Cantonese farce, and soap operas with traditional legends of anti-heroes to explore the "'anonymous and unsung heroes' of the Pearl River Delta or 'the factory of the world.'" She has also created a virtual city in *Second Life*, featured at the Venice Biennale, which I will discuss in Part Three.

10 To experience the constantly frustrated and roadblocked Web surfing experience as a person in China would, you can download Toby Linegruber, Aram Bartholl, Evan Roth's Firefox add-on, *The China Channel*. It redirects your browsing experience behind the Great Firewall. Download it at http://chinachannel.fffff.at/.

11 This ad is posted on YouTube (http://www.youtube.com/watch?v=D9n24aP6lkA), but YouTube itself is not viewable in China. Chinese must watch the ad at Sina's own site (http://bit.ly/qC6Cpa).

12 Even films that are granted permission to be shown are sometimes censored, as in the recent case of *Pirates of the Caribbean: The World's End* (2008). Some violent scenes were cut, making the action hard to follow. Chow Yun-Fat's scenes at the end of the film were trimmed, as they were deemed vilifying and humiliating to the Chinese.

13 Alberro cites Marshall McLuhan and Quentin Fiore's famous statement from *The Medium is the Massage* (1967):

> Xerography – every man's brain-picker – heralds the times of instant publishing. Anybody can now become both author and publisher. Take any books on any subject and custom-make your own book by simply Xeroxing a chapter from this one, a chapter from that one – instant steal! (quoted in Alberro, 2003, 130).

14 In the former Eastern bloc, they are called "dilemma actions" (http://www.
nytimes.com/2011/07/15/world/europe/15belarus.html).

15 Witness how Twitter protest helped bring down part of the empire of the
previously untouchable newspaper tycoon Robert Murdoch after his tabloid,
the *News of the World,* was discovered to have hacked into the cell phones
of a murdered girl, Members of Parliament and the Prime Minister, and was
revealed to have spent hundreds of thousands of pounds bribing police.

Disturbance (action + event)

1 The original video was posted to the Web in pre-YouTube days, but its
popularity soared when it was later reposted on YouTube.

Capture/leakage (performance + documentation)

1 There are several numbers and versions of this photographic series.

2 For a more in-depth discussion of Second Front and simulated acts of
terrorism, see my article, "All The Rage: The Digital Body and Deadly Play in
the Age of the Suicide Bomber" in *There's A Gunman on Campus: Tragedy
and Terror at Virginia Tech* Ben Agger and Timothy W. Luke, ed. New York:
Rowman & Littlefield, 2008, pp. 215–30.

3 See Coco Fusco's interview with Obadike at http://www.blacknetart.com/
coco.html.

4 A recent case is soon to go to trial in an Australian copyright suit that has
as its prime evidence a WikiLeaks cable. Accused of not doing enough
to enforce copyright by an Australian agency, WikiLeaks revealed that the
accusation actually originated with the Motion Picture Association of America
through the MPA, its international agent.

> The MPA, "does not want that fact to be broadcasted," the 2008
> communiqué from then Ambassador Robert D. McCallum Jr. explained.
> "MPAA prefers that its leading role not be made public," the summary of
> the case added, to dodge the impression that it is "just Hollywood 'bullying
> some poor little Australian ISP'."

> This revelation, along with earlier leaks, once again raises a disturbing
> question. How far are the US State Department and US-based content
> industries intruding into the IP affairs of other countries – particularly
> members of The Commonwealth? (Lasar, 2011).

5 Luksch is not the only artist to use live capture as a way of creating media
either. Many artists also use the audience as content. See in particular,
Christiane Paul's discussion of *WAXWEB,* Mark Amerika's *Filmtext* and Nick
Crowe's *Discrete Packets* (Paul, 2008, 102–9).

6 As I have written elsewhere (in *artwomen.org* and in *Extensions*), Paul Virilio states in *Open Sky* that in our multimodal, mediatized world, the old concepts of objectivity and subjectivity or no longer adequate. He posits the concept of the *trajective subject*, a subject that exists in a state of perpetual motion between subjective and objective points of view (Virilio, 1997, 24).

7 For an extensive exploration of Eija-Liisa ahtila's works see Alberro's article (2008).

8 These masks are different faces that can be adopted at will for protective purposes. The ability to "pass" makes Mestiza and women of color less susceptible to the violence of their oppressors, Anzaldua argues, but also involves a denial of their origins and situation (1990, xv).

Part Two: From karaoke culture to vernacular video

1 At a cost of $30 to $40 a head at his July 2011 concert in Toronto.

2 Her most recent works are bronzes and remix Georgia O'Keeffe's antelope skulls as bronzed sculptures and adapt ritual body masks into wall hangings in metal.

3 A sympatico project was undertaken by the MTAA Collective (a Brooklyn-based conceptualist net.art duo, Mark River and T. Whid) with *Ten Digital Readymades* (2000).

4 In 2007, Eric Kluitenberg said:

> The mass-media are about to dissolve into a sea of hypermedial fragments, transforming into a multitude of hybrids and singularities (does anybody still know what television actually is these days?). This inevitably invites a radical fragmentation of 'the public'. This is a process that has at long taken hold of the informational societies. The current explosion of self-publication in countless weblogs, on community websites, self-video portals, in on-line diaries, web fora and a plethora of individual websites is only the visible sign of an undercurrent that was already for many years transforming 'the public' into an amalgamation of increasingly unrelated subjectivities and singular interest groups. What can be witnessed today is the rise of swarm publics – highly unstable constellations of temporary alliance, creating a public sphere in constant flux – globally mediated flash mobs that never meet – fuelled by sentiment and affect – escaping fixed capture. The face of "radical mediocrity."

This is what Kluitenberg calls the "Society of the Unspectacular" (Kluitenberg, 2007). On the surface, at least, it bears an uneasy resemblance to the excess of sensation Aldous Huxley warned us we would be underwhelmed with in *Fahrenheit 451*. The artistry instead, I argue, is not in the end disposable or virtual product, but in the actions and effects it engenders.

5 Like *Silent Star Wars* (http://www.openculture.com/2011/07/

star_wars_as_silent_film.html) or *The Empire Strikes Back (1950)* (whoiseyevan).

6 Clipart can be used for higher ends too. Witness, for instance, Oliver Laric's *787 Cliparts.* A profoundly smooth animation of 787 pre-existing copyright free GIF images, Laric brings these unrelated images together seamlessly to create a poetic dance of fluid motion. He calls it "a lightning-fast flipbook of people, costumes and cultures." Flipbooks were one of the earliest forms of animation; a series of stop motion images drawn or printed on subsequent pages of a book were made dynamic when a reader flipped through the pages.

7 Current standards are 32 and 64 bit.

8 An even more novel use of YouTube is as a source material for studying cognition and mapping dreams or the imagination. In 2011, a team of scientists at the University of California at Berkeley created a 3D model of the brain and mapped how the voxels, the pixels of brain space, interpret and visualize shapes and movement. They then studied the blood flow in the brain as scientists watched movie trailers while inside an MRI machine. Having fed 18 million random seconds of YouTube clips into their program, the computer selects the 100 most similar clips and layers them together to reconstruct the original vision. In retrospect, it was only a matter of time before someone decided to use YouTube as data, but using it to map the imagination is a particularly brilliant stroke. No doubt before too long similar technology will be used to map how we 'see' sound (Kottke: http://kottke.org/11/09/an-actual-working-mind-probe). Here is what the subject were shown and how the computer reconstructed it: http://www.youtube.com/watch?v=nsjDnYxJ0bo&feature=player_embedded (Nishimoto).

'Aberrant decoding' and atactical aesthetics

1 At a subsequent screening in 1969, Cornell chose a rose-colored filter for the projector instead (http://en.wikipedia.org/wiki/Rose_Hobart_(film).

2 The only reference I could find to Nestor Amaral's album says that *Holiday in Brazil* was a 1957 rerelease of an earlier album called simply *Brazil* (http://stax-o-wax.blogspot.com/2008/10/nestor-amaral-and-his-continentals.html). I have been unable to find the release date of the original, and so do not know if this is the music that Cornell played for the original screening in 1936 or if this was a new choice for the 1969 showing. Amaral seems to have been popular on Hollywood soundtracks in the 1940s and 1950s, as in the 1941 *Weekend in Havana* and in the 1948 film *Romance on the High Seas* where he sang a duet of "It's Magic" with Doris Day. It seems possible that Amaral could have been recording as early as 1936.

3 According to *Wikipedia:*

> Vidding began in 1975, when Kandy Fong synced Star Trek stills on a slide projector with music from a cassette player. She performed her

vids with live cutting at fan conventions, which continue to be one of the main venues for vid-watching. When home videocassette tape recorders became available in the mid-1970s, vidders began producing live-action vids that were recorded onto media that could be shown at fan conventions and further distributed to fans. Substantial technical and artistic skill were required to cut vids together, often requiring footage from multiple VCRs to be placed on the same tape, with the added challenge of exact timing. Typical vids could take 6–8 hours to produce, and more elaborate ones could take substantially longer. Vidders, predominately women, passed this knowledge on to each other.

With the rise of digital media, greater bandwidth, and the widespread availability of free, albeit basic, video creating/editing software such as iMovie and Windows Movie Maker the skill level required for vidding has been reduced and the number of distribution outlets has increased. As a result, both the number of vidders and the number of accessible vids has skyrocketed. However, because of concerns that the outside community won't understand the vids and the context of vidding as well as some copyright and intellectual property concerns, many of the most experienced vidders do not make their vids readily-accessible on public venues such as YouTube, although this is changing (http://en.wikipedia.org/wiki/Vidding).

4 Compare this to Jim Newell's (BadLipReading's) *"Rick Perry" – A BLR Soundbite*, which translates a campaign speech into psychobabble.

5 Strelka was one of the dogs that the Russians sent into space.

6 Another example (presumably analog, as 'new media' in China generally means analog video) is by the so-called 'Father of Chinese video art', Zhang Peili's *Actor's Lines* (2002). It appropriates, and echoes an edited passage of pre-Cultural Revolution classic propaganda film, *Soldiers under Neon Lights* by Wang Pei (1964). Zhang's film replays and replays a patriotic soldier's attempts to win the confidence of a young man over and over until it becomes a rhythmic, stuttering, meaningless revisitation. Zhang's 2003 *Last Words* plays a stream of death scenes from 1950s revolutionary Chinese films across two screens. On one screen the Communist heroes die, on the other they are miraculously reborn in the same scene reversed. Instead of dying, they wake into life. This 15-minute endless loop renders life and death meaningless. They enact "an incisive pastiche of a well-known phrase from the revolutionary era – 'heroes are immortal'" ("Zhang Peili", *Synthetic Reality*).

Google Empire: smart art, intelligent agents

1 When designing Pivot, Gary Flake at Microsoft set out to design a tool that would allow us to view the Web as a web rather than as a series of pages.

2 Barret Lyon's Opte Project seeks to make the Internet 'visible.' Started in October 2003, it is an open source code map of the Internet.

3 Inxight's *StarTree*, a visualization engine for extremely complex datasets,

performs a similar server-side function for government contractors' or software vendors' data.

4 In "Autonomous Interface Agents," Henry Lieberman, designer of the Web-browsing agent called *Letizia*, defines 'interface agents' and 'autonomous agents' as different kinds of organisms (http://lieber.www. media.mit.edu/people/lieber/Lieberary/Letizia/AIA/AIA.html). Agents are conversational, meaning that we cue them through give-and-take behavior to perform, whereas autonomous agents are always running and pursue their own desires in the background regardless of what we are doing. I do not intend to dispute this, but autonomous agents still engage with us at the level of the interface, the point at which we meet the machine.

5 In fact, *Daria* is so eager to please that she sounds more like someone in the business of selling sexual services than digital art.

6 A variety of ancient experimental agent projects are showcased at the MIT MediaLab Software Agents Group (http://agents.media.mit.edu/). One of the more interesting ones, now defunct, was called *Your Memory, Connected* by James Chao-Ming Teng, Edward Shen, Edward Lieberman, and Patti Maes. *Your Memory, Connected* was a computational memory access program that created a collage from images it harvested from Flickr. Other agents include *BUZZwatch* that keeps track of trends, *Footprints* that follows other users' breadcrumbs, and *Letizia* that searches ahead in anticipation of a browser's needs. A more familiar agent may be *Ask Jeeves,* the self-proclaimed world's first Internet butler (http://www.ask.com).

Real time

1 There used to be a third, *Slit Scan Camera* by Funner Labs, but it is no longer available.

Part Three: Creative cannibalism and digital anthropophagy

1 Yifan Wong's work *Light and Shadow*, an interactive façade, also mistranslates traditional Chinese shadow puppetry through digital media into the context of Qinqiang opera. Working with a gesture and media system, the interactor's shadow is translated into the silhouttes of traditional opera characters by four infrared cameras that fire as the coordinates are updated within the system as fast as 120 frames per second. Initially the work only has one character and the second actor is a female opera performer hired from the local company. When completed (it is still in the prototype stage), the system will support two characters.

'Productive mistranslation' (China and Pakistan)

1 Of the more than 100 DVDs I sampled while I was in Pakistan in 2009, only three were of a lower quality than factory-released commercial ones and three were damaged or would not play at all. While the quality was in most cases identical, they did not all have the standard commercial interface. Their packaging, however, was often riddled with spelling mistakes or comical mistranslations, or cut-and-pasted material from other films, and the disk itself is always plain and unlabeled unless it was an omnibus edition.

2 According to Alexandra Crosby (2008, 8):

> The style of this video, and much Indonesian video art, comes from this immersion in rapidly developing technologies and their applications to popular culture. Particularly influential has been the marketing and distribution of 'low-fi' pornography. The residue of this style is evident in *Bermain* with its voyeuristic feel and disregard for technical fluency. The more deliberate aspect of the production is the digital storytelling methods, designed through workshops focusing, in this case, on the ever-important issue of space in Jakarta.

3 Shezad Dawood, for one, has moved from taking Western movie posters and "re-imagining them with a post-colonial shift" to producing feature films.

4 Compare this to Carmen Karasic and Critical Art Ensemble's *FloodNet*, discussed earlier.

5 *The Rising Tide's* Website is viewable at http://www.mohattapalacemuseum. com/THE%20RISING%20TIDE.html.

6 The trend has viral attributes too as it spreads to the rest of the art world.

7 This irony is sometimes lost on Western collectors who powered the sale of political pop on the international market. Marianne Brouwer says that:

> in Western eyes, modern art WAS the West. One of the consequences of this point of view was the voyeuristic appreciation of "Eastern" dissident art through Western eyes. The notion that freedom of expression was the exclusive, historical achievement of Western society since the Enlightenment, led to the inverse reasoning that all art from repressed countries like Russia or China had to be "dissident" in order to be "authentic" (Brouwer, 2004, 5).

8 Britta Erickson claims that Wang Gongxin was the first Chinese artist to "master computer manipulation of video" (2005, 114), but Wang Jianwei's digital work seems to have preceded Wang Gongxin's by as much as two years.

9 That scene from *Titanic* jumps to a copycat scene in a Korean or Taiwanese film on the back of a truck.

10 I have been unable to confirm who is correct, but in *Video Art* Michael Rush says that "Image and Phenomena" was held in *1994* (not 1996) in Hangzhou (2007, 143).

Bibliography

"'5 Million Dollars 1 Terabyte': Pirated Data Displayed at Art." *Huffington Post*, September 7, 2011. Available at www.huffingtonpost.com/2011/09/06/ five-million- dollars-of-s_n_951150.html>.

88seamonkeys. "Indonesia Art Activism and Rock 'n Roll." *YouTube*, November 14, 2006. Web, September 23, 2011: www.youtube.com/ watch?v=m4qJsUOYms0>.

Aaron, Michele. *Spectatorship: The Power of Looking On.* London: Wallflower Press, 2007.

Aczel, Amir D. *The Mystery of the Aleph.* New York: Simon & Schuster, 2001.

Adorno, Theodor, and Max Horkheimer, "The Culture Industry: Enlightenment as Mass Deception," In *Dialectic of Enlightenment.* Stanford, CA: Stanford University Press, 2002. [1944].

Agamben, Giorgio. "Difference and Repetition: On Guy Debord's Films." In *Art and the Moving Image: A Critical Reader*, ed. Tanya Leighton. London: Tate, 2008, pp. 328–33.

Akira, Tatehata. "Why Cubism? (2006)." In *Contemporary Art in Asia: A Critical Reader*, ed. Melissa Chiu and Benjamin Genocchio. Cambridge, MA, and London: MIT Press, 2011, pp. 107–15.

Alberro, Alexander. "At the Threshold of Art as Information." In *Recording Conceptual Art: Early Interviews with Barry, Huebler, Kaltenbach, LeWitt, Morris, Oppenheim, Siegelaub, Smithson, Weiner*, ed. Alexander Alberro and Patricia Novell. Berkeley and Los Angeles: University of California Press, 2001, pp. 1–15.

—*Conceptual Art and the Politics of Publicity.* Cambridge, MA: MIT, 2003.

—"The Way Out Is the Way In." In *Art after Conceptual Art*, ed. Alexander Alberro and Sabeth Buchmann. Cambridge, MA, and London: MIT Press, 2006, pp. 13–26.

—"The Gap Between Film and Installation Art." In *Art and the Moving Image: A Critical Reader*, ed. Tanya Leighton. London: Tate, 2008, pp. 423–9.

"Alterazioni Video" (2009/10). *Piemonte Share.* Web, July 23, 2011: www. toshare.it/?page_id=3031&lang=en>.

Anzaldua, Gloria. *Borderlands/La Frontera: The New Mestiza.* San Francisco, CA: Aunt Lute, 1987.

—"Haciendo Caras, Una Entrada." Introduction. In *Making Face, Making Soul: Haciendo Caras*, ed. Gloria Anzaldua. San Francisco, CA: Aunt Lute, 1990, pp. xv–xxviii.

AP. "Russia's Illicit Industry Booming – Business – World Business – Msnbc.com." *Msnbc.com.* July 7, 2006. Web, December 9, 2011: http:// www.msnbc.msn.com/id/13617606/ns/business-world_business/t/ russias-illicit-industry-booming/.

Arcade Fire: The Wilderness Downtown. Director Chris Milk, 2010. Web,
    September 2010.
Arcangel, Cory. "Super Mario Movie." *YouTube*, September 27, 2010. Web: www.
    youtube.com/watch?v=_2vjnBF_oXY.
Armitage, John. "Beyond Postmodernism: Paul Virilo's Hypermodern Cultural
    Theory." *CTheory.* November 15, 2002. Web, June 12, 2011: www.ctheory.
    net/articles.aspx?id=133.
Art Bienniale. "Art Biennale 2011 – Christian Marclay Interview." *YouTube*, June
    6, 2011. Web: www.youtube.com/watch?v=IPz55aeSLL0.
ArtLeaks. Web: October 5, 2011: www.art-leaks.org/.
ArtSlant. "White Box: Exhibition Detail: Sept. 13th to Oct. 9th, 2011, Anton S.
    Kandinsky, Ai Weiwei" (2011). Web, September 20, 2011: www.artslant.com/
    ny/events/show/177365-china-ism-ii-democracy-or-economy.
Aufderheide, Patricia. "How Documentary Filmmakers Overcame Their Fear of
    Quoting and Learned to Employ Fair Use: a Tale of Scholarship in Action."
    International Journal of Communications 1 (2007): 26–36.
Augé, Marc. *Non-places: An Introduction to Supermodernity* (2nd edn). London:
    Verso, 2008.
Auslander, Philip. *Liveness: Performance in a Mediatized Culture.* London and
    New York: Routledge, 1999.
Bach, Christen. *Bear Untitled – D.O. Edit.* YouTube, April 20, 2010. Web: www.
    youtube.com/watch?v=cdBZohX4FnM.
Bagchi, Jeebesh, Lawrence Liang, Ravi Sundaram, and Sudhir
    Krishnaswamy, eds. *Contested Commons/Trespassing Publics:
    A Public Record.* Delhi, India: Sarai Programme, 2005. *Welcome
    to Sarai 2014 S A R A I.* The Sarai Programme, November 2005.
    Web, January 1, 2011: www.sarai.net/publications/occasional/
    contested-commons-trespassing-publics-a-public-record.
Ball, Lucille, director. "Lucy and Harpo Marx." *I Love Lucy*, May 9, 1955.
Banowetz, Michael, and Noah Sadona. *This Aborted Earth: The Quest
    Begins.* YouTube, March 27, 2010. Web, June 3, 2011: www.youtube.com/
    watch?v=G9geI52zIxI.
Bard, Perry. "When Film and Database Collide." In *Video Vortext Reader II:
    Moving Images Beyond YouTube*, ed. Geert Lovink and Rachel Somers Miles.
    Amsterdam: Institute of Network Cultures, 2011, pp. 322–9. Web, March
    10, 2011: www.networkcultures.org/wpmu/portal/publications/inc-readers/
    video-vortex-ii/.
Barthes, Roland. "The Death of the Author." *Aspen* 5+6 (1967). UbuWeb, Web,
    September 23, 2011: www.ubu.com/aspen/aspen5and6/index.html.
Basilico, Stephen. "The Editor." In *Cut: Film as Found Object in Contemporary Video*,
    ed. Stefano Basilico. Milwaukee, WI: Milwaukee Art Museum, 2004, pp. 29–45.
Bassnett, Susan. *Translation Studies*, revised edn. New York: Routledge, 1991.
Baudrillaud, Jean. "Ectasy of Communication." In *The Anti-Aesthetic*, ed. Hal
    Foster. New York: The Bay Press, 1983, pp. 126–34.
—"Beyond the Vanishing Point of Art." In *Post-Pop Art*, ed. Paul Taylor.
    Cambridge, MA: MIT Press, 1988.
—*Simulacra and Simulation*, trans. Shiela Faria Glaser. Michigan: University of
    Michigan Press, 1994 [1991].

—*Spirit of Terrorism*, trans. Chris Turner. London and New York: Verso, 2002.

"Beam Up The Bass." *Addictive* TV. 2011. Web. 05 Mar. 2012. <http://www.addictive.com/videos>.

Beccaria, Marcella, and Candice Breitz. *Candice Breitz*. Milano: Skira Editore, 2005.

Benjamin, Walter. "Work of Art in the Age of Mechanical Reproduction." In *Illuminations*, ed. Hannah Arendt, trans. Harry Zohn. New York: Schocken, 1968, pp. 83–109.

Bennett, Andrew. *The Author (The New Critical Idiom)*. New York: Routledge, 2005.

Berger, Kevin. "Classical Music Waltzes with Digital Media." *L.A. Times Online*, Web, August 7, 2011: www.latimes.com/entertainment/news/music/la-ca-classical- technology-20110807,0,7795377.story.

Bermain – Playing. Director Melisa Deyatri. *EngageMedia*, July 17, 2007. Web: www.engagemedia.org/Members/okvideo/videos/playing-xvid.avi/view.

Berners-Lee, Tim. "What the Semantic Web Can Represent." *W3.org*. 1998. Web, April 15, 2004: www.w3.org/DesignIssues/RDFnot.html.

Bey, Hakim. *Temporary Autonomous Zone*. Automedia, 1985. Web, June 14, 2011: www.hermetic.com/bey/taz_cont.html.

Beyoncé. "Single Ladies (Put A Ring On It)." YouTube, October 2, 2009. Web, September 27, 2011: www.youtube.com/watch?v=4m1EFMoRFvY.

Bhabha, Homi K. *The Location of Culture*. London: Routledge, 1994.

Bigelow, Alan. *Brainstrips. Electronic Literature Organization Anthology, Volume 2*, 2009. Web: www.collection.eliterature.org/2/works/bigelow_brainstrips.html.

Binghui, Huangfu. "Long March Space: Interview of Wang Jianwei." *Longmarchspace.com*. Sydney, Australia, May 2004. Web, September 23, 2011: www.longmarchspace.com/artist/list_28_words.html?locale=en_US.

Bishop, Claire. "Antagonism and Relational Aesthetics." *October* 110 (Fall 2004): 51–80.

Bittanti, Matteo. "Game Art: Josh Bricker's 'Post Newtonianism'" (2010)." *Gamesciences: Art in the Age of Computer Games*, October 31, 2010. Web, June 14, 2011: www.gamescenes.org/2010/10/game-art-josh-brickers-post-newtonianism-2010.html.

Blackmore, Susan. *The Meme Machine*. Oxford: Oxford University Press, 1999.

Blocker, Jane. *What the Body Cost: Desire, History and Performance*. Minneapolis: University of Minnesota Press, 2004.

Blumlinger, Christa. "The Art of the Possible: Notes about Some Installations by Harun Farocki." In *Art and the Moving Image: A Critical Reader*, ed. Tanya Leighton. London: Tate, 2008, pp. 273–82.

Bogdan, Deanne. "Music, McLuhan, Modality: Musical Experience from 'Extreme Occasion' to 'Alchemy'." *MediaTropes* 1 (2008): 71–101. Web: www.mediatropes.com.

Bolter, J. David, and Richard Grusin. *Remediation: Understanding New Media*. Cambridge, MA: MIT Press, 2000.

Boon, Marcus. *In Praise of Copying*. Cambridge, MA: Harvard University Press, 2010.

Bourriaud, Nicolas. *Postproduction: Culture as Screenplay: How Art Reprograms the World*. New York: Lukas & Sternberg, 2002a.

—*Relational Aesthetics*. *Umelec*, April 2002b. Web, May 26, 2008: www.divus.cz/umelec/article_page.php?item=932.

—"Relational Aesthetics." In *Participation: Documents of Contemporary Art*, ed. Claire Bishop. London and New York: Whitechapel/MIT Press, 2006, pp. 160–71.

Boyle, James. "Fencing Off Ideas: Enclosure and the Disappearance of the Public Domain." In *CODE: Collaborative Ownership and the Digital Economy*, ed. Rishab Aiyer Ghosh. Cambridge, MA, and London: MIT Press, 2005, pp. 235–58.

Bradshaw, Peter. "Christian Marclay's *The Clock*: A Masterpiece of Our Times." *Guardian Online*, April 7, 2011. Web, June 6, 2011: www.guardian.co.uk/film/filmblog/2011/apr/07/christian-marclay-the-clock.

—"Is This the Greatest YouTube Clip of All Time?" *Guardian Online*, October 17 2007. Web, June 6, 2011: www.guardian.co.uk/film/filmblog/2007/oct/17/isthisthegreatestyoutubec.

Brandt, Daniel. "PageRank: Google's Original Sin." *Nettime*, August 27, 2002: www.google-watch.org/pagerank.html.

Bray, Christopher. "*The Beach Beneath the Street* by McKenzie Wark: Review." *Observer Online*, August 28, 2011: www.guardian.co.uk/books/2011/aug/28/beach-beneath-street-wark-review.

Bray, Tim. *Visual Net*. Computer software. Web: www.antarctica.net/products/visual_net/.

Breitz, Candice. *Soliloquies*. Web, June 13, 2011: www.candicebreitz.net/.

Brokeback Mountain. Director. Ang Lee. Producer, Diana Ossana and James Schamus. By Diana Ossana and Larry McMurtry. Universal Home Entertainment, 2005 (DVD).

Brouwer, Marianne. "Making Noise in the West, To Attack in the East." *Synthetic Reality*, Catalog. Hong Kong: Timezone 8, 2004, pp. 3–10.

Burgess, Jean. "'All Your Chocolate Rain Are Belong to Us'? Viral Video, YouTube the Dynamics of Participatory Culture." In *Video Vortex Reader: Responses to YouTube*, ed. Geert Lovink and Sabine Niederer. Amsterdam: Institute of Network Cultures, 2008, pp. 101–9.

Butt, Danny. "Local Knowledge and New Media." In *Place: Local Knowledge and New Media Practice*, ed. Danny Butt, Jon Bywater, and Nova Paul. Newcastle: Cambridge University Press.

Caesar, Ray. "A Dangerous Inclination." Corey Helford Gallery, New York, September 2011. Web: www.arrestedmotion.com/2011/09/teaser-ray-caesar-a-dangerous-inclination-corey-helford/.

Cammaer, Gerda Johanna, Curator. *Brisk Collages and Bricolages*. Catalog. Halifax, NS: Mount St. Vincent Art Gallery, 2005.

Cao, Fei. *Caofei.com*. Web: www.caofei.com/biography.html.

—"Shadow Life." *New York Times Online Video Library*. Web: www.video.nytimes.com/video/2011/06/02/arts/design/100000000846503/cao-feis-shadow-life.html.

Cardiff, Janet, and George Bures Miller. *Ghost Machine*. Berlin, 1995. Web, May 31, 2010: www.cardiffmiller.com/artworks/walks/ghostmachine.html.

Caronia, Antonio. "Shadows of Video Art: The New Era of Quite (*sic*) Images." *Digimag* 58 (October 2010). Web: www.digicult.it/digimag/article.asp?id=1907.

Carr, David. "Mourning Old Media's Decline." *New York Times Online*, October

28, 2008. Web, September 23, 2011: www.nytimes.com/2008/10/29/business/media/29carr.html?fta=y.

Carriero-Granados, Rebecca. "Russian Constructivist Art and Its Influence on Film." *Russian Constructionist Art and Film 66*, ed. Spearmint. Web, January 3, 2009: www.hubpages.com/hub/Russian-Constructivist-Art-and-Film.

Case, Sue-Ellen. *Domain-Matrix: Performing Lesbian at the End of Print Culture*. Bloomington and Indianapolis: Indiana University Press, 1996.

Cassell, David. "Hacktivism in the Cyberstreets." *AlterNet*, May 30, 2000. Web, September 27, 2011: www.alternet.org/story/9223.

Catacchio, Chad. "Today Finland Officially Becomes First Nation to Make Broadband a Legal Right." *The Next Web – EU*, July 1, 2010. Web, July 1, 2010: www.thenextweb.com/eu/2010/07/01/today-finland-officially-becomes-first-nation-to-make-broadband-a-legal-right/.

Cavell, Richard. "Gould, McLuhan, and the Fate of the Acoustic." *Glenn Gould* 10.1 (2004): 44–7.

Center for Social Media. "Fair Use." May 25, 2008. Web, December 9, 2011: www.centerforsocialmedia.org/fair-use.

Chao, Loretta. "Chinese Authorities Trying to Come to Terms with Weibo?" *Wall Street Journal*, August 4, 2011. Web, August 8, 2011: www.blogs.wsj.com/chinarealtime/2011/08/04/chinese-authorities-trying-to-come-to-terms-with-weibo/tab/print/.

China Tracy: i.Mirror, Part 1. Directors Cao Fei and China Tracy. YouTube, June 3, 2007. Web, July 5, 2011: www.youtube.com/watch?v=5vcR7OkzHkI.

China Tracy: i.Mirror, Part 2. Directors Cao Fei and China Tracy. YouTube, June 4, 2007. Web, July 5, 2011: www.youtube.com/watch?v=jD8yZhMWkw0.

China Tracy: i.Mirror, Part 3. Directors Cao Fei and China Tracy. YouTube, June 3, 2007. Web, July 5, 2011: www.youtube.com/watch?v=zB-ILJlnWEE.

Chiu, Melissa, and Benjamin Genocchio. "Introduction: What Is Asian Contemporary Art? Mapping an Evolving Discourse." In *Contemporary Art in Asia: A Critical Reader*. Cambridge, MA, and London: MIT Press, 2011, pp. 1–12.

"Christian Marclay: *The Clock*." *BBC News at 10:00*, November 4, 2010. Web, June 6, 2011: www.youtube.com/watch?v=Y8svkK7d7sY.

Cizek, Katerina. *Highrise: The Story So Far*, 2010. Web, September 30, 2011: www.highrise.nfb.ca/index.php/the-story-so-far.

Clay, Andrew. "Blocking, Tracking, and Monetizing: YouTube Copyright Control and the *Downfall* Parodies." In *Video Vortext Reader II: Moving Images Beyond YouTube*, ed. Geert Lovink and Rachel Summers. Amsterdam: Institute of Network Cultures, 2011, March 8, 2011. Web, March 10 2011: www.networkcultures.org/wpmu/weblog/2011/03/08/video-vortex-reader-ii-moving-images-beyond-youtube-new-inc-publication/.

Cohen, David. "When the Rule Ruled." Web log post, *Artcritical.com*, March 3, 2005. Web, September 18, 2011: www.artcritical.com/DavidCohen/SUN90.htm.

Collage D'Hollywood. Director Richard Kerr. Canada 2008, 8 min, 35mm film (in Bricolage).

Condon, Brody. *Karma Physics < Elvis*. Video Documentation of Game Experience, 2004. Web, September 23, 2011: www.tmpspace.com/?portfolio=karmaphysics.

—*Death Animations, 2007–2008*. Documentation of

Perfomance. Web, September 24, 2011: www.tmpspace. com/?portfolio=death-animations-2007-2008.

Connerton, Paul. *How Societies Remember*. Cambridge: University of Cambridge Press, 1989.

Conradi, and Christensen. *Stateless Plug-in*. Computer software, 2011. Web, October 5, 2011: www.statelessplugin.net/.

Coupland, Douglas. "Why McLuhan's Chilling Vision Still Matters Today." Editorial, *Guardian*, July 21, 2011: www.guardian.co.uk/commentisfree/2011/jul/20/marshall-mcluhan- chilling-vision?INTCMP=SRCH.

Coverley, Merlin. *Psychogeography*. Harpenden: Pocket Essentials, 2006.

Cowan,T.L. "DaynaMcLeod'sFeministCelluloidFantasies,QueerRevisionary: MakingItBetter." *NoMorePotlucks*Motive#16(August–September2011). Web,September10,2011:www.nomorepotlucks.org/article/motive-no-16/dayna-mcleod's-feminist-celluloid-fantasies-queer-revisionary-making-it-better.

Cramer, Florian. "The Fiction of the Creative Industries." *Nettime-l*, 24 Sept. 2011. Published in *Open* by the Dutch Foundation for Art and the Public Space (Stichting Kunst en Openbare Ruimte/SKOR). The Dutch language publication is here: www.skor.nl/nl/site/item/open-noodnummer-over-de-nieuwe-politiek-van-cultuur.

Crandall, Jordan. "Jordan Crandall's Heat Seeking. (2000). Documentation." *Jordan Crandall*, 2000. Web, February 2, 2011: www.jordancrandall.com/heatseeking/heatsK.html.

—"Precision + Guided + Seeing." *CTheory.net* 1000 Days of Theory (2006). *CTheory.net*, October 1, 2006. Web, September 15, 2011: www.ctheory.net/articles.aspx?id=502.

"The Criminalization of Copyright Infringement in the Digital Era." *Harvard Law Review* 112.7 (1999): 1705–1722. *JSTOR*. Web, September 25, 2011: www.jstor.org/stable/1342415.

Crimp, Douglas. "Sherrie Levine: 'After Ed Weston,' 1980." *October* 15 (Winter 1980). Web, March 3, 2011: www.artnotart.com/sherrielevine/crimpweston.html.

Critical Art Ensemble. "Recombinant Theatre and Digital Resistance." *The Drama Review* 44.4 (T168) (Winter 2000): 151–66.

—*Digital Resistance*, 2001. Web, January 2011: www.critical-art.net/books/digital/.

Crosby, Alexandra. "Collaborative Cultural Practices of Activist Networks in Contemporary Indonesia." *17th Biennial Conference of the Asian Studies Association of Australia*. Conference Proceedings, Melbourne, July 1–3, 2008. Web, September 23, 2011: www.arts.monash.edu.au/mai/asaa/alexandracrosby.pdf.

Darkvx. "Where the Hell Is Matt? WoW Version." YouTube, August 16, 2008. Web, July 21, 2011: www.youtube.com/watch?v=mUULMN_Y-ps&feature=related.

DataCell. "News: Visa and Mastercard Transactions Rejected." *DataCell – Coolest Datacenter on the Planet*, August 12, 2010. Web, October 5, 2011: www.datacell.com/news/.

Dawes, Brendan. "Slit Scan: Don't Look Now ~ Documentation. Brendan

Dawes." *Brendan Dawes' Don't Look Now*, 2004. Web, January 2, 2011: www.brendandawes.com/project/slit-scan-dont-look-now/.

De Botton, Alain. "Christian Marclay's Video Installation *The Clock*." *The BBC 3: Culture Show*, November 11, 2010. Web, September 13, 2011: www.youtube. com/watch?v=rB3CgEnxnYY.

De Certeau, Michel. *The Practice of Everyday Life*, trans. Steven Rendall. Berkely, Los Angeles, CA, and London: University of California Press, 1984.

—*Heterologies: Discourses on the Other*, trans. Brian Massumi. Minneapolis: University of Minneapolis Press, 1986. Theory of History and Literature Series, Vol. 17.

"De Digitale Stad (DDS) <<The Digital City>> ." *Medien Kunst Netz/Media Art Net*. Web: www.medienkunstnetz.de/works/digitale-stad/?desc=full.

De Kluitenberg, Eric. "Remix – Society of the Unspectacular." *Nettime-l*, June 10, 2007. Web, June 10, 2007: www.nettime.org/Lists-Archives/nettime-l-0706/msg00010.html.

De Vietri, Gabrielle. *Captcha*, February 2011. Web, July 23, 2011: www.vimeo. com/18383130.

Debord, Guy. "Methods of *Détournement*." *Les Lèvres Nues* #8 (May 1956). Web, June 10, 2011: www.library.nothingness.org/articles/SI/en/display/3.

—*Society of the Spectacle*, trans. Ken Knabb. London: Rebel Press, 1983 [1967].

—"Theory of the Dérive." *Situationist International Anthology*, ed. Ken Knabb, trans. Ken Knabb. Bureau of Public Secrets, 1981.

Delete Yourself. Director Tasman Richardson, Canada 2004. One-minute videotape.

Dickinson, Rod, and Steve Rushton. "Who, What, Where, When, Why and How (2009). Documentation." *Rod Dickinson – Artist*, July 11, 2009. Web, July 9, 2011: www.roddickinson.net/pages/whowhatwhere/project-synopsis.php.

Dirtyfoxhead. "Hitler Pissed About Too Many Downfall Parodies." YouTube, April 4, 2010. Web, July 20, 2011: www.youtube.com/watch?v=ChEXf91j2pU.

Disney Interactive Studios. *Epic Mickey*. Burbank, CA: Disney, 2010 (Nintendo Wii Game software).

Dodes, Rachel. "'Life in a Day' Director Aims to Elevate YouTube Videos Into Art." *Speakeasy: Wall Street Journal*, July 22, 2011. Web, September 3, 2011: www.blogs.wsj.com/speakeasy/2011/07/22/ life-in-a-day-director-kevin-macdonald-aims-to-elevate-youtube-videos-into-art/?mod=google_news_blog.

Dodge, Martin. "Mapping How People Use a Website." *Mappa.Mundi Magazine*, March 30, 2004: www.mappa.mundi.net/maps/maps_022/.

Dowling, Siobhan. "Pirate Party Snatches Seats in Berlin State Election." *Guardian Online*, September 18, 2011: www.guardian.co.uk/world/2011/ sep/18/pirate-party-germany-berlin-election.

Downfall (der Untergang). Director Oliver Hirschbiegel, 2004 (DVD).

Duck Soup. Director Leo McCarey, 1933.

Dûrsteler, Juan C. *KartOO. Inf@Vis! The Digital Magazine of InfoVis.net*, August 19, 2002. Web, December 8, 2003: www.kartoo.net/e/en/artpresse/2002/ press_aout/infovis.htm.

Eco, Umberto. "Innovation and Repetition: Between Modern and Post-Modern Aesthetics." *Daedalus* 114.4, *The Moving Image* (Fall 1985): 161–84.

Edmunds, Michael. "Flip-Side Overlap: The Medium Is the Music." *MediaTropes*

1 (2008): 102–12. Web, October 7, 2010: www.mediatropes.com/index.php/ Mediatropes/article/viewFile/3007/1486.

Egoyan, Atom. *81/2 Screens*. Installation. September 2010.

"Egypt Flips Domain Name System Kill Switch." *Domain Names Australia & Australian Domain Name Registration Services*, January 29, 2011. Web, June 13, 2011: www.domainregistration.com.au/news/egypt-domain-name-system. php.

Elahi, Hasan. "Tracking Transience V2.0." *Hasan Elahi ~ Tracking Transience*. 2002. Web, September 15, 2011: www.elahi.umd.edu/track/.

"Electronic Sit-in in Solidarity with People's Protests in Iran against Rigged 2009 Presidential Elections." *Electronic Sit-in in Solidarity with People's Protests in Iran against Rigged 2009 Presidential Elections*. 21 June 2009. Web. 05 Mar. 2012. <http://iran2009election.opinionware.net/>.

Elsaesser, Thomas. "Introduction: Harun Farocki." *Senses of Cinema* 21 (2002). Web, July 23, 2011: www.sensesofcinema.com/2002/21/ farocki_intro/.

The Empire Strikes Back (1950). Director Whoiseyevan. YouTube, May 28, 2010. Web, August 12, 2011: www.youtube.com/watch?v=KmTpOQrqoO0&feature =related.

Endo, Takumi, and Shinya Masuyama. "Open Source Art from DIVVY/dual: Interview." *PinMag Tokyo*, September 21, 2006. Web, December 10, 2011: www.pingmag.jp/2006/09/21/open-source-art-from-divvydual/.

—*TypeTrace. Software*. Computer software. *TypeTrace: Record Your Writing*, December 13, 2007. Web, December 10, 2011: www.typetrace.jp/.

—*TypeTrace*. Tokyo: DIVVY/Dual, 2006.

Endo, Takumi. "Interview: Takumi Endo" by Galerie CIANT, May 25, 2008. Web: www.prague.tv/articles/art-and-culture/takumi-endo-interview.

Englehardt, Tom. "Reading in an Age of Depression." *The Nation*, December 18, 2008. Web, January 5, 2009: www.thenation.com/article/ reading-age-depression.

English, Leona. "Third Space: Contested Space, Identity, and International Adult Education." 2002. Web, May 25, 2008: www.oise.utoronto.ca/CASAE/ cnf2002/2002_Papers/english_l2002w.pdf.

"Epic Fail." *LOLCats*, November 7, 2007. Web, August 13, 2011: www. icanhascheezburger.com/2007/11/07/epic-fail-2/.

Erickson, Britta. ""Here? Or There?: A History." *Here? or There? Wang Gongxin and Lin Tianmiao*. China: Timezone 8, 2005. 110–14.

Faceless. Director Manu Luksch. AmbientTV.net, 2007 (DVD).

Fannin, Rebecca. "From Made in China to Invented in China: 'Chinov8!'" *Forbes. com*, September 22, 2011. Web, September 22, 2011: www.forbes.com/ sites/rebeccafannin/2011/09/22/from-made-in-china-to-invented-in-china- chinov8/?partner=technology_newsletter.

Fast, Omar. *CNN Concatenated*. YouTube. Web, June 2, 2011: www.youtube. com/watch?v=RCD3IxCZpsM.

The Father Divine Project. Web, June 2, 2011: www.fatherdivine.wordpress.com/.

"Feng Mengbo. Interview." Online interview, September 18, 2011.

Flusser, Vilém. *Towards a Philosophy of Photography*, trans. Anthony Mathews. London: Reaktion, 2000.

Foucault, Michel. "What Is an Author?" In *The Foucault Reader*, ed. Paul Rabinow. London: Penguin, 1986 [1969]. 141–60.
—*The Order of Things*. New York: Vintage, 1994 [1970].
Franke, Anselm. "The Ghosts Crossing the Smokescreen: On the Position of Wang Jianwei's Work in the Emerging Debate." Longmarchspace.com, March 4, 2011. Web, September 23, 2011: www.longmarchspace.com/artist/list_28_words.html?locale=en_US.
Frohne, Ursula. "Dissolution of the Frame: Immersion and Participation in Video Installations." In *Art and the Moving Image: A Critical Reader*, ed. Tanya Leighton. London: Tate, 2008, pp. 355–70.
Fry, Ben. *Anemone*. Web, August 25, 2010: www.benfry.com/anemone/.
Fuller, Matthew. "Commonality, Pixel Property, Education: As If." Web, September 25, 2011: www.year01.com/plunder/essay.html.
—"Towards an Evil Media Studies." *SPC.org*, March 2007. Web, September 29, 2011: www.spc.org/fuller/texts/towardsevil/.
Funkhouser, Chris. "Le(s) Mange Texte(s): Creative Cannibalism and Digital Poetry." *E-Poetry*, 2007. Web, April 2011: www.epoetry2007.net/english/funkhouseruk.org. Unpublished conference paper presented at Université Paris 8.
—"Cannabalistic Tendencies in Digital Poetry: Recent Observations & Personal Practices." *Chris Funkhouser ~ Online Publications*. Web, June 13, 2011: www.web.njit.edu/~funkhous/online.html.
Fusco, Coco. "All Too Real: Tale of an Online Black Sale. Fusco Interviews Keith Townsend Obadike." *Blacknetart.com*, September 24, 2001. Web, June 12, 2011: www.blacknetart.com/coco.html.
Gallantlabucb. "Nishimoto,etal. 2011. Reconstruction Mpeg." YouTube, September 21, 2011. Web, September 23, 2011: www.youtube.com/watch?v=nsjDnYxJ0bo.
Galloway, Alexander R. *Protocol: How Control Exists After Decentralization*. Cambridge, MA: MIT Press, 2004.
Garcia, David, and Geert Lovink. "The ABC of Tactical Media." *Nettime-l*, May 16, 1997. Web, January 2011: www.fdm2.ucsc.edu/~mhaughwo/170a/w10/media/manifestos/TheABCofTacticalMedia.pdf.
Gellman, Dara. *Alien Kisses*. Three-minute video installation. 1997.
Gibson, Steve. "Grand Theft Bicycle. Documentation." *GTB*, 2007. Web, October 5, 2011: www.grandtheftbicycle.com/.
—"Trans-Art Exhibition DAW, Xi'an Academy of Fine Art. Documentation." In *Digital Arts Weeks: The Meeting Point Between Art Technology and Culture*. July 2010. Web, July 2011: www.digitalartweeks.ethz.ch/web/DAW10/Trans-Art.
Gillespie, Tarleton. *Wired Shut: Copyright and the Shape of Digital Culture*. Cambridge, MA: MIT Press, 2007.
Gmooney. "Christian Marclay: Mini Documentary." *TIVO*, November 7, 2006. Web, June 6, 2011: www.youtube.com/watch?v=4yqM3dAqTzs.
Godard, Barbara. "Theorizing Feminist Discourse/Translation." *Tessera* 6 (March 1989): 42–53.
Golan, Doran. "An Informal Catalogue of Slit-Scan Video Artworks and Research – Golan Levin and Collaborators." *Flong – Interactive Art by Golan Levin and Collaborators*, July 17, 2010. Web, January 2, 2011: www.flong.com/texts/lists/slit_scan/.

Greene, Rachel. *Internet Art*. London: Thames and Hudson, 2004.

Guertin, Carolyn. "Glistening Bodies: Cyberfeminism, Interactivity and Slattery's Collaborinth." *Artwomen.org*, July 30, 2002. Web: www.artwomen.org/cyberfems/guertin/guertin1.htm.

—"Wanderlust: The Kinesthetic Browser in Cyberfeminist Space." *Extensions: The Online Journal of Embodiment and Technology* 3 (2007). Web: www.performancestudies.ucla.edu/extensionsjournal/guertin.htm.

—"All the Rage: The Digital Body and Deadly Play in the Age of the Suicide Bomber." In *There's a Gunman on Campus: Tragedy and Terror at Virginia Tech*, ed. Ben Agger and Timothy W. Luke. New York: Rowman & Littlefield, 2008a, pp. 215–30.

—"Handholding, Remixing and The Instant Replay: New Narratives in a Post-Narrative World." In *A Companion to Digital Literary Studies*, ed. Susan Schreibman and Ray Siemens. Oxford: Blackwell, 2008b. Web: www.digitalhumanities.org/companionDLS/.

Hamilton, Anita. "Top 10 Viral Videos (2008): Where the Hell Is Matt?" *TIME Magazine Online*, November 3, 2008. Web, July 1, 2011: www.time.com/time/specials/packages/article/0,28804,1855948_1864281_1864282,00.html.

Haraway, Donna. "The Cyborg Manifesto." In *Simians, Cyborgs and Women*. New York: Routledge, 1991 [1984], pp. 149–81.

Harding, Matt. "Where the Hell Is Matt? (2005)." YouTube, June 24, 2006. Web, July 1, 2011: www.youtube.com/watch?v=7WmMcqp670s.

—"Where the Hell Is Matt? (2006)." YouTube, June 20, 2006. Web, July 1, 2011: www.youtube.com/watch?v=bNF_P281Uu4.

—"Where the Hell Is Matt? (2008)." YouTube, June 20, 2008. Web, July 1, 2011: www.youtube.com/watch?v=zlfKdbWwruY.

—"Matt Confesses: Where the Hell Is Matt? Video an 'Elaborate Hoax' – YouTube." YouTube, January 2, 2009. Web, July 1, 2011: www.youtube.com/watch?v=ogcqFaNbah4.

Hardt, Michael, and Antonio Negri. *Empire*. Cambridge, MA: Harvard University Press, 2000.

—*Multitude: War and Democracy in the Age of Empire*. New York and London: Penguin, 2004.

Harley, Ross Rudesch. "Cultural Modulation and the Zero Originality Clause of Remix Culture in Australian Contemporary Art," ed. Tofts McCrea, Darren McCrea, and Christian McCrea. *The Fibreculture Journal 15: 2009 Remix* (2009). Web, August 29, 2011: www.fifteen.fibreculturejournal.org/fcj-100-cultural-modulation-and-the-zero-originality- clause-of-remix-culture-in-australian-contemporary-ar/.

Harmon, Amy. "'Hacktivists' of All Persuasions Take Their Struggle to the Web." *New York Times Online*. October 31, 1998. Web, September 27, 2011: www.nytimes.com/library/tech/98/10/biztech/articles/31hack.html.

Harold, Christine. *OurSpace: Resisting the Corporate Control of Culture*. Minneapolis: University of Minnesota Press, 2007.

Hartigan, Brian. "Loving Lucy." *TV Guide* (November 10–16, 2001): 12–16.

Harvey, Sylvia. *May '68 and Film Culture*. London: British Film Institute, 1980.

Hayles, N. Katherine. "Materiality of Informatics." *Configurations* 1.1 (1993): 147–70.

—*How We Became Posthuman*. Chicago: University of Chicago Press, 1999.

Helft, Miguel, and Motoko Rich. "Google Settles Suit Over Book-Scanning." *New York Times Online*, October 28, 2009. Web: www.nytimes.com/2008/10/29/technology/internet/29google.html?fta=y.

Heller, Michael. *The Gridlock Economy: How Too Much Ownership Wrecks Markets, Stops Innovation, and Costs Lives*. New York: Basic Books, 2008.

Hitlerrantsparodies. "Hitler Rants about the Downfall Parodies Being Blocked." YouTube, April 5, 2010. Web, July 20, 2011: www.youtube.com/watch?v=5fAQKa8rU_4.

—"Breaking Her Own Rules: Dayna McLeod Talks TV Remix." In *Vague Terrain: Digital Art/Culture/Technology*, August 8, 2011. Web, September 10, 2011: www.vagueterrain.net/content/2011/08/breaking-her-own-rules-dayna-mcleod-talks-tv-remix.

Hogan, Mel. "Performing Politics: An Interview with Dayna McLeod." In *Art Threat: Culture and Politics*, September 18, 2007. Web, September 10, 2011: www.artthreat.net/2007/09/performing-politics-an-interview-with-dayna-mcleod/.

Houston, Tom. "Julian Assange: Facebook Is Appalling Spy Machine." *Huffington Post*, May 2, 2010. Web, September 5, 2011: www.huffingtonpost.com/2011/05/02/julian-assange-facebook-is-spy-machine_n_856313.htm.

Howard, Jen. "Public Wins With White Spaces." *Free Press*, November 4, 2008. Web: www.freepress.net/node/45595.

Hoy, Amy, and Thomas Fuchs. *Twistori*. 25 May 2008. Web: www.twistori.com.

Huang Du. "Interview: Microcosm ~ A Modern Allegory." In *Miao Xiaochun: Macromania*. Ludwig Museum Catalog. Hong Kong: Timezone 8, 2010.

Humphries, Matthew. "Aaron Swartz Spent Months Stealing Data from MIT, Now Facing 35 Years in Prison." *Geek.com*, July 19, 2011. Web, July 20, 2011: www.geek.com/articles/geek-pick/aaron-swartz-spent-months-stealing-data-from-mit-now-facing-35-years-in-prison-20110719/.

Hung, Wu. "A Case of Being 'Contemporary': Conditions, Spheres and Narratives of Contemporary Chinese Art (2008)." In *Contemporary Art in Asia: A Critical Reader*, ed. Melissa Chiu and Benjamin Genocchio. Cambridge, MA, and London: MIT Press, 2011, pp. 391–410.

Huot, Claire. *China's New Cultural Scene: A Handbook of Changes*. Durham, NC, and London: Duke University Press, 2000.

Huyghe, Pierre. *Remake*. Two-channel video installation, (1994–5).

Indonesia Art Activism and Rock 'n Roll. Director Jamie Nicolai and Charlie Hillsmith. SBS, 2004.

The Infinity Lab. "The Infinity Lab Does The Getty." *The Infinity Lab*. Web, July 13, 2011: www.theinfinitylab.com/getty.html. See also: www.katherinesweetman.com/the_getty.html.

I/O/D 4. *Web Stalker* (1997–8). Web, February 26, 2005: www.backspace.org/iod/.

Ippolito, Jon. "Why Art Should Be Free." *Thoughtmesh*, August 2002. Web, May 24, 2008: www.three.org/ippolito/writing/why_art_should_be_free/#.

Iyengar, Jayanthi. "Intellectual Property Piracy Rocks China Boat." *Asia Times Online*, September 16, 2004. Web, May 26, 2008: www.atimes.com/atimes/China/FI16Ad07.html.

Jana, Reena. "Is It Art, or Memorex?" *Wired*, May 2005. Web, Septembe 18, 2011: www.wired.com/print/culture/lifestyle/news/2001/05/43902.

Jaschko, Susanne, "Space-Time Correlations Focussed in Film Objects and Interactive Video." FILE Symposium, Sao Paulo, Brazil, 2002. Online: www.sujaschko.de/en/research/pr1/spa.html.

Jenkins, Henry. "If It Doesn't Spread, It's Dead: Media Viruses and Memes." *Web log post*, February 11, 2009. Web, July 28, 2011: www.henryjenkins.org/2009/02/if_it_doesnt_spread_its_dead_p.html.

"Jim Campbell." *EArt: New Technologies and Contemporary Art*. Web, September 20, 2011: www.fondation-langlois.org/e-art/e/jim-campbell.html.

Jinjua, Dai. "Immediacy, Parody, and Image in the Mirror: Is There a Postmodern Scene in Bejing?", trans. Jing M. Wang. In *Multiple Modernities: Cinemas and Popular Media in Transcultural East Asiain*, ed. Jenny Kwok Wah Lau. Philadelphia, PA: Temple University Press, 2003, pp. 151–66.

"Jodi (Art Collective)." *Wikipedia*, July 14, 2010. Web, August 25, 2010: www.en.wikipedia.org/wiki/Jodi_(art_collective).

Jon. "Open Source Art from DIVVY/dual." *PingMag*, September 21, 2006. Web, May 25, 2008: www.pingmag.jp/2006/09/21/open-source-art-from-divvydual/.

Jones, Amelia. *Body Art: Performing the Subject*. Minneapolis: University of Minnesota Press, 1998.

Jorgensen, John. "Twitter Turns Down $500 Million in Facebook Stock." *TinyComb*, November 24, 2008. Web, January 3, 2009: www.tinycomb.com/2008/11/24/twitter-turns-down-500-million-in-stock-from-facebook/.

Joris, Pierre. "On the Seamlessly Nomadic Future of Collage." In *Cutting Across Media: Appropriation Art, Interventionist Collage and Copyright Law*, ed. Kembrew McLeod and Rudolf Kuenzli. Durham, NC, and London: Duke University Press, 2011, pp. 185–98.

Kaizen, William. "Live on Tape: Video, Liveness and the Immediate." In *Art and the Moving Image: A Critical Reader*, ed. Tanya Leighton. London: Tate, 2008, pp. 257–72.

Kandinsky, Anton S. "Kandinsky About China-ism," April 28, 2010. Web, August 1, 2011: www.antonkandinsky.com/literature/chinaism.htm.

Kaprow, Allan. "Untitled Guidelines for Happenings (1965)." In *Multimedia: from Wagner to Virtual Reality*, by Randall Packer and Ken Jordan. New York: Norton, 2002, pp. 307–14.

Karasick, Adeena. "Lingual Ladies (Put a Frame on It)." YouTube, February 26, 2011. Web, September 27, 2011: www.youtube.com/watch?v=q7MjU609Lwo.

Kildall, Scott. *Kildall.com* (2006). Web: www.kildall.com/.

—*Paradise Ahead* (2007). Second Life Performance and C-Print. Documentation: www.kildall.com/artwork/2007/paradise_ahead/index.html.

—"Paradise Ahead." *Kildall.com* (2007). Web, August 5, 2007: www.kildall.com/artwork/2007/paradise_ahead/index.html.

Kirschenbaum, Matt. "A White Paper in Information." *Digital Arts and Culture*, December 6, 1998. Web, January 2, 2002: www.jefferson.village.virginia.edu/~mgk3k/white/index.html.

Kline, David, and Dan Burnstein. *Blog!* New York: CDS, 2005.

Kluitenberg, Eric. "Transfiguration of the Avant-Garde: The Negative Dialectics of the Net." *Nettime-l*, January 2, 2002: www.nettime.org.
—"Remix: Society of the Unspectacular." *Nettime-l*, June 10, 2007: www.nettime.org/Lists-Archives/nettime-l-0706/msg00010.html.
Kohout, Martin. *Moonwalk* (2008). Web, September 24 2011, www.martinkohout.com/moonwalk.
Kottke, Jason. "An Actual Working Mind Probe." *Blog: Kottke.org*, September 23, 2011: www.kottke.org/11/09/an-actual-working-mind-probe.
Krauss, Rosalind. "Video: The Aesthetics of Narcissism." In *Art and the Moving Image: A Critical Reader*, ed. Tanya Leighton. London: Tate, 2008, pp. 208–19.
Kretschmann, Bryce. "Auspice: Sqomb." YouTube, June 25, 2010. Web, September 24, 2011: www.youtube.com/watch?v=ijv7hYVxMdY.
Kruks, Sonia. *Retrieving Experience: Subjectivity and Recognition in Feminist Politics*. Ithaca, NY: Cornell University Press, 2001.
Krulwich, Robert. "Extreme Tidying Up: Ursus Wehrli." Web log post, *NPR*, September 12, 2011/12: www.npr.org/blogs/krulwich/2011/09/12/140394807/extreme-tidying-up?sc=fb.
Kuo, Kaiser. "Intellectual Property: The Benefits of Piracy." *Danwei*, August 12, 2006. Web, May 26, 2009: www.danwei.org/intellectual_property/the_benefits_of_piracy_by_kais.php.
Lange, David. "Reimagining the Public Domain." *Law and Contemporary Problems* 66 (Winter/Spring 2003): 463–83. Web: www.law.duke.edu/shell/cite.pl?66+Law+&+Contemp.+Probs.+463+%28WinterSpring+2003%29.
Langheinrich, Ulf, and Kurt Henschlager. "Granular Synthesis Pol." YouTube, 1998. Web, September 30, 2011: www.youtube.com/watch?v=Wpp7u-bGXw8. See also www.medienkunstnetz.de/works/pol.
Lanks, Belinda. "Animated GIFs Capture Stanley Kubrick's Most Immortal Scenes." *Co.Design*, August 31, 2011. Web, August 31, 2011: www.fastcodesign.com/1664907/animated-gifs-capture-stanley-kubricks-most-immortal-shots.
Laric, Oliver. "Versions, 2010." YouTube. Web, July 23, 2011: www.youtube.com/watch?v=X5OwM5zA-cQ.
Larkin, Brian. "Pirate Infrastructures." In *Structures of Participation in Digital Culture*, ed. Joe Karaganis. New York: Social Science Research Council, 2007, pp. 75–84. Web: www.ssrc.org/publications/view/6A130B0A-234A-DE11-AFAC-001CC477EC70/.
Lasar, Matthew. "WikiLeaks: MPAA Behind Aussie ISP Lawsuit (But Don't Tell Anybody)." *Ars Technica*, Septembe 2, 2011: www.arstechnica.com/tech-policy/news/2011/09/wikileaks-mpaa-behind-aussie-isp-lawsuit-but-dont-tell-anybody.ars.
Lau, Jenny Kwok Wah. "Introduction." In *Multiple Modernities: Cinemas and Popular Media in Transcultural East Asiain*, ed. Jenny Kwok Wah Lau. Philadelphia, PA: Temple University Press, 2003, pp. 1–12.
Laurel, Brenda. *Computers as Theatre*. Reading, MA: Addison-Wesley, 1997.
Lee, Timothy B. "Shock, Awe: British Government Agrees That Copyright Has Gone Too Far." *Ars Technica*, August 2011. Web, October 5, 2011: www.arstechnica.com/tech-policy/news/2011/08/the-british-government-has-endorsed.ars.

Leiberman, Henry. "Letizia." Web, May 5, 2011: www.lieber.vwww.media.mit. edu/people/lieber/Lieberary/Letizia/AIA/AIA.html.

Lessig, Lawrence. *Code: And Other Laws of Cyberspace.* New York: Basic Books, 1999.

—*Free Culture: How Big Media Uses Technology and Law to Lock Down Culture and Control Creativity.* New York: Penguin, 2004. Web: www. free-culture.cc.

Levin, Golan, and Iris Dressler. "An Informal Catalogue of Slit-Scan Video Artworks and Research – Golan Levin and Collaborators." In *Flong – Interactive Art by Golan Levin and Collaborators*, July 17, 2010. Web, February 15, 2011: www.flong.com/texts/lists/slit_scan/.

Li, Pi. "'Synthetic Reality': A Historical Note." In *Synthetic Reality.* Hong Kong: Timezone 8, 2004, pp. 15–17.

Lichty, Patrick. "The Issue of Remediation in SL Art." *Empyre-l*, August 5, 2007: www.voyd.com/texts/LichtyEmpyreSLMissive3RemediationinSLArt.pdf.

Lillemose, Jacob. "Conceptual Transformations of Art: From Dematerialisation of the Object to Immateriality in Networks." In *Curating Immateriality: The Work of the Curator in the Age of Network Systems*, ed. Joasia Krysa. Autonomedia, 2006. DATA Browser Ser. #03. Web: www.kurator.org/ publications/curating-immateriality/.

Lima, Manuel. *Visual Complexity: Mapping Patterns of Information.* New York: Princeton Architectural, 2011.

Linegruber, Tony, Aram Bartholl, and Evan Roth. *The China Channel.* Firefox Add-On, 2008. Web: www.chinachannel.fffff.at/.

Lovink, Geert. "Blogging: The Nihilist Impulse." *Eurozine*, February 1, 2007: www.eurozine.com/articles/2007-01-02-lovink-en.html#footNoteNUM5. Based on a lecture given at the Berlin Institute of Advanced Study, the Wissenschaftsklleg, Berlin, March 27, 2006.

—"The Art of Watching Databases: Introduction to the Video Vortex Reader." In *Video Vortex Reader: Responses to YouTube*, ed. Geert Lovink and Sabine Niederer. Amsterdam: Institute of Network Cultures, 2008, pp. 9–12.

—*Zero Comments: Blogging and Critical Internet Culture.* New York and Oxon: Routledge, Taylor & Francis, 2008b.

Luksch, Manu. "Manifesto for CCTV Filmmakers." *AmbientTV.net*, 2006: www. ambienttv.net/content/?q=dpamanifesto.

Lunenfeld, Peter. *The Secret War between Downloading and Uploading: Tales of the ????.*

Malik, Atteqa. "Interview." Via Skype: Karachi, Pakistan, September 28, 2011.

Man with a Movie Camera. Director Perry Bard, July 23, 2010: www.dziga. perrybard.net/.

Manovich, Lev. "Remix and Remixability." *Nettime*, November 6, 2005. Web, January 29, 2006.

—"The Poetics of Augmented Space." *Visual Communication* 5.2 (2006): 219–40.

Mantel, Gustaf. "If We Don't, Remember Me." October 2010. Web, August 31, 2011: www.iwdrm.tumblr.com/.

Marclay, Christian. "Christian Marclay." Interview by Alain De Botton. YouTube. Originally aired on BBC, *Culture Show*, November 11, 2010: www.youtube. com/watch?v=rB3CgEnxnYY.

—"Christian Marclay." Interview by Venice Biennale. YouTube, June 6, 2011: www.youtube.com/watch?v=IPz55aeSLL0.

Marker, Chris. *Silent Movie*. Video installation, 2005.

Marshall, M. "Remix Is a Cultural Right, Lessig Says." *Virginia Law*, Fall 2004. Web, May 26, 2008: www.law.virginia.edu/html/news/2004_fall/lessig.htm.

Mason, Matt. *The Pirate's Dilemma: How Youth Culture Is Reinventing Capitalism*. New York: Free Press, 2008.

MastersofHumility. "Hitler is Fed up with All the Hitler Rants!" YouTube, March 15, 2009. Web, July 20. 2011: www.youtube.com/watch?v=7vMUvgce_5s.

The Matrix. Dir. Andy Wachowski and Larry Wachowski. Warner Home Video, 1999. DVD.

Mattes, Eva and Franco. "Eva and Franco Mattes Reenacting Joseph Beuys' '7000 Oaks'." *Rhizome*, March 20, 2007. Web, June 12, 2011: www.rhizome.org/discuss/view/25069/.

—"New Sculpture by Cattelan Turns Out to Be An Art Prank." December 7, 2010. Web, August 17, 2011: www.0100101110101101.org/home/catt/.

McCarty, Willard. "'Knowing True Things by What Their Mockeries Be:' Modelling in the Humanities." *TEXT Technology* 12.1 (2003): 43–58.

McClymont, Alistair. "Art for Art's Sake." Nine-monitor installation (2007). Web, June 24, 2011: www.alistairmcclymont.com/artwork/art-arts-sake.

McGrath, Charles. "'Dancing,' a Near-perfect Piece of Internet Art." *New York Times Online*, July 8, 2008. Web, July 1, 2011: www.nytimes.com/2008/07/08/business/worldbusiness/08iht-dancing.4.14336375.html.

McLaren, Malcolm. "Authentic Creativity vs. Karaoke Culture." *TED Talks: Best of the Web*, October 2009. Web, May 30, 2011: www.ted.com/talks/malcolm_mclaren_authentic_creativity_vs_karaoke_culture.html.

McLaren, Peter. "Multiculturalism and the Post-modern Critique: Toward a Pedagogy of Resistance and Transformation." In *Between Borders: Pedagogy and the Politics of Cultural Studies*, ed. Henri Giroux and Peter McLaren. New York: Routledge, 1994, pp. 193–222.

McLeod, Dayna. *That's Right Diana Barry – You Needed Me*. Quicktime Movie (6:04), *Daynarama.com*, November 29, 2008. Web, July 2, 2011: www.daynarama.com/HTML2/Anne.html.

McLeod, Kembrew, and Rudolf E. Kuenzli. "I collage Therefore I am: An Intro to Cutting Across Media." In *Cutting across Media: Appropriation Art, Interventionist Collage, and Copyright Law*, ed. Kembrew McLeod and Rudolf E. Kuenzli. Durham, NC, and London: Duke University Press, 2011, pp. 1–23.

McLuhan, Marshall. *Understanding Media: The Extensions of Man*. Toronto: Signet, 1964.

—"The Angelic Discarnate Man of the Electric Age." Interview by Patrick Peyton, November 14, 1971. Web, October 8, 2010: www.i-boy.com/weblog/2009/03/angelic-discarnate- man-of-electric-age.html.

—*Letters of Marshall McLuhan*, ed. Matie Molinaro, Corrine McLuhan, and William Toye. Toronto: Oxford University Press, 1987.

—"Address at Vision 65." In *Essential McLuhan*, ed. Eric McLuhan and Frank Zingrone. Toronto: Anansi, 1995, pp. 208–21.

—"Art as Survival in the Electric Age." In *Understanding Me: Lectures and*

Interviews, ed. Stephanie McLuhan and David Staines. Toronto: McClelland and Stewart, 2004 [1973].

McLuhan, Marshall, and Quentin Fiore. *Medium Is the Massage: An Inventory of Effects.* Toronto: Penguin, 2003 [1967].

McVeigh, Karen. "Wall Street Protesters: Over-educated, Under-employed and Angry." *Guardian Online*, September 19, 2011. Web, September 25, 2011: www.guardian.co.uk/world/2011/sep/19/wall-street-protesters-angry.

"Media Art Net|Sollfrank, Cornelia: Female Extension." *Medien Kunst Netz|Homepage*: August 11, 2011: www.medienkunstnetz.de/works/female-extension/.

MediaShed. Web:.www.mediashed.org/.

Meek, Allen. "Indigenous Virtualities." In *Place: Local Knowledge and New Media Practice*, ed. Danny Butt, Jon Bywater, and Nova Paul. Newcastle: Cambridge Scholars Publishing, 2008, pp. 20–34.

Meeting of Two Queens: Encuentro Entre Dos Reinas. Director Cecilia Barriga (1991).

Mendoza, Antonio. "State of the Union." YouTube, September 8, 2006. Web, September 29, 2011: www.youtube.com/watch?v=ChAmKGEZrkI.

Metelerkamp, Peter. "Questioning Realism: Bazin and Photography." 2001. Accessed March 1, 2011: www.metelerkamp.com/writing/on-photographic-realism/.

Mfishe4. "Star Wars: The Empire Strikes Brokeback." YouTube, October 31, 2006. Web, September 20, 2011: www.youtube.com/watch?v=iuV9lXU5rDI.

"Miao Xiaochun. "Interview." Online, September 20, 2011.

Miller, Paul D. (a.k.a. DJ Spooky, That Subliminal Kid). *Rhythm Science*. Amsterdam and New York: MIT Mediaworks, 2004.

—*Sound Unbound*. New York: MIT Press, 2008.

Mitchem, Matthew. "Video Social: Complex Parasitical Media." In *Video Vortex Reader: Responses to YouTube*, ed. Geert Lovink and Sabine Niederer. Amsterdam: Institute of Network Cultures, 2008, pp. 273–82.

Modern Language Association. *MLA Style Manual and Guide to Scholarly Publishing* (3rd edn). New York: Modern Language Association of America, 2008.

Moser, David. "Media Schizophrenia in China." *Danwei*, August 7, 2006. Web, May 26, 2008: www.danwei.org/media_and_advertising/media_schizophrenia_in_china_b.php.

Mosireen. 2011. Web, October 5, 2011: www.mosireen.org/.

Mosley, Alonzo. "100 Movies, 100 Quotes, 100 Numbers." YouTube, February 19, 2007. Web, June 6, 2011: www.youtube.com/watch?v=FExqG6LdWHU.

Mouffe, Chantal. "Artistic Activism and Agonistic Spaces." *Art and Research* 1.2 (2007): 1–5. Web, September 27, 2011: www.artandresearch.org.uk/v1n2/mouffe.html.

"Moviebarcode." *Moviebarcode*. Web, January 29, 2011: www.moviebarcode.tumblr.com/.

Moynihan, Colin. "80 Arrests as Protesters March in New York." *New York Times*, September 25, 2011a: 18.

—"Occupying, and Now Publishing, Too." *New York Times (blog)*, October 1, 2011b. Web, October 2, 2011: www.cityroom.blogs.nytimes.com/2011/10/01/occupying-and-now-publishing-too/.

MTAA Collective (Mark River and T. Whid). *Ten Digital Readymades* (2000). Web, June 26, 2011: www.mteww.com/readymade/index.html.

Mueller, Kurt, and Chelsea Beck. "Press Release: NEW SCULPTURE BY CATTELAN TURNS OUT TO BE AN ART PRANK." *Inman Gallery*, December 7, 2010. Web, July 13, 2011: www.inmangallery.com/artists/WEASEL/2010_WEASEL_press_updated.pdf.

Mulji, Huma. "Interview." Via Skype: Lahore, Pakistan, September 19, 2011.

Mulji, Huma, and Shilpa Gupta, Curators. *Aar-Paar Text* (2004).

Mundrawala, Asma. *Eid Attraction*. Computer Print, 2000: www.members.tripod.com/aarpaar2/00asma.htm.

Münter, Ulrike. "Chinesische Gegenwartskunst: Where Next for Mao? Shi Xinning's Director's Notes on Chinese History." Online Catalog, September 2007. Web, June 26, 2011: www.chinesische-gegenwartskunst.de/pages/katalogtexte/wohin-mit-mao-en.php.

Muresan, Ciprian. *Un Chien Andalou*. Silent digital video, 3D animation, one minute, 2004. Web, September 25, 2011: www.vimeo.com/9048229.

Murray, Janet. *Hamlet on the Holodeck*. Cambridge: MIT Press, 1997.

Nakashima, Ellen. "Expanded Powers to Search Travelers at Border Detailed." *Washington Post Online*, September 23, 2008. Web, November 5, 2008: www.washingtonpost.com/wp-dyn/content/article/2008/09/22/AR2008092202843.html.

NationMaster. "Crime Statistics: Piracy Rate (Most Recent) by Country." *NationMaster*, 2005. Web, June 2, 2008: www.nationmaster.com/graph/cri_sof_pir_rat-crime-software-piracy-rate.

Navas, Eduardo. "Remix Defined." *Remix Theory*. Web, September 30, 2011: www.remixtheory.net/?page_id=3.

Negroponte, Nicholas. *Being Digital*. New York: Alfred A. Knopf, 1995.

Nelson, Cary. "AAUP Legislastive Alert: Border Searches of Electronic Materials" (October 14, 2008). Online newsletter.

Newell, Jim (BadLipReading). *Rick Perry – A BLR Soundbyte*. YouTube, September 18, 2011: www.youtube.com/watch?v=BhDhDRvHaGs.

Ni, Haifeng. "The Power of Culture: What Is Synthetic Reality?" *Synthetic Reality*. 2002. Web, September 21 2011: www.powerofculture.nl/uk/specials/synthetic_reality/faq.html.

Ni Haifeng, and Pauline Yao. "Conversation between Ni Haifeng and Pauline Yao." Web, September 1, 2011: www.haifeng.home.xs4all.nl/h-text-para%20interview.htm.

Noah Kalina Everyday. Director Noah Kalina. YouTube – *Broadcast Yourself*, August 27, 2006. Web, October 2, 2011: www.youtube.com/watch?v=6B26asyGKDo.

Nothing Compares To You ~ House Mashup. Director Dayna McLeod. Performer Hugh Laurie. Daynarama, July 27, 2011. Web, August 3, 2011: www.daynarama.com/HTML2/House.html.

Ohyearvideos. "Brokeback Batman." YouTube, December 4, 2006. Web, September 10, 2011: www.youtube.com/watch?v=RnSIIgyAX9k.

Onetrickdog. "Brokeback of the Ring." YouTube, February 27, 2007. Web, September 10, 2011: www.youtube.com/watch?v=tgt-BiFiBek.

Out My Window. Director Katerina Cizek (2010). Web, September 30, 2011: www.interactive.nfb.ca/#/outmywindow.

Owens, Craig. "The Discourse of Others: Feminists and Postmodernism." *Art/ Not Art.* Web: www.artnotart.com/sherrielevine/owensexcerpt.html.

Paik, Nam June. *Good Morning, Mr. Orwell* (1984). *Media Art Net.* Web: www.medienkunstnetz.de/works/goog-morning/.

Panhans-Bühler, Ursula. "Double Interface: Miao Xiaochun's *RESTART.*" *Miao Xiaochun: Macromania*, Ludwig Museum Catalog. Hong Kong: Timezone 8, 2010.

Paul, Christine. *Digital Art*, Vol. 2E. New York and London: Thames and Hudson, 2008.

Pennings, Mark. "Relational Aesthectics and Critical Culture." In *Proceedings Transforming Aesthetics*. Art Gallery of New South Wales, 2005, pp. 1–9. Web: www.eprints.qut.edu.au/archive/00006996/01/6996.pdf.

Peretti, Jonah. "Notes on Contagious Media." In *Structures of Participation in Digital Culture*, ed. Joe Karaganis. New York: Social Science Research Council, 2007. Web: www.ssrc.org/publications/view/6A130B0A-234A-DE11-AFAC-001CC477EC70/.

Perreault, John. "Rirkrit Tiravanija: Fear Eats the Soul." *Artopia Blog, ArtsJournal. com*, April 2011. Web, August 13, 2011: www.artsjournal.com/artopia/2011/04/rirkrit_tiravanija_fear_eats_t_1.html.

Phelan, Peggy. *Unmarked: The Politics of Performance*. London and New York: Routledge, 1993.

Philippe Pareeno and Pierre Huyghe: No Ghost Just a Shell (1999–2002). Kunsthalle Zürich Catalog, August 24 to October 27, 2002. Web, September 27, 2011: www.mmparis.com/noghost.html.

Pi, Li. "Chinese Contemporary Video Art." *Third Text* 23.3 (2009): 303–7.

Pirates of the Amazon. "Documentation," December 1, 2008. Web, June 21, 2011: www.pirates-of-the-amazon.com/.

Podcast Network. "Twittories." *Twistori*, May 24, 2008. Web: www.twistori.com/#.

Pontin, Jason. "From Many Tweets, One Loud Voice on the Internet." *New York Times Online*, April 22, 2007. Web, May 25, 2008: www.nytimes.com/2007/04/22/business/yourmoney/22stream.html.

Puranen, Jorma. "Imaginary Homecoming//1999." In *Appropriation*, ed. David Evans. London: Whitechapel, 2009, pp. 135–6. Documents of Contemporary Art.

Quaranta, Domenico. "The Real Thing: Interview with Oliver Laric." *Art Pulse Magazine*. Web, July 23, 2011: www.artpulsemagazine.com/the-real-thing-interview-with-oliver-laric/.

Rai2121. "Harry Potter and the Brokeback Goblet." YouTube, March 8, 2006. Web, September 10, 2011: www.youtube.com/watch?v=e9D0veHTxh0.

Raley, Rita. "Reveal Codes: Hypertext and Performance." *Postmodern Culture* 12.1 (2001). Web: www.muse.jhu.edu/journals/pmc/v012/12.1raley.html.

—*Tactical Media*. Minneapolis: University of Minnesota Press, 2009.

Ramos-Velasquez, Vanessa. *Quiet Revolution*. Web, February 2, 2011: www.quietrevolution.me/Vanessa_Ramos_Velasquez.html.

Raqs Media Collective. "Pacific Parables." In *PLACE: Local Knowledge and New Media*. Newcastle, UK: Cambridge Scholars, 2008, pp. 9–19. Web: www. raqsmediacollective.net/texts9.html.

Ravetto-Biagioli, Kriss. "Shadowed by Images: Rafael Lozano-Hemmer and the Art of Surveillance." *Representations* 111.1 (Summer 2010): 121–43.

"The Real Thing: Interview with Oliver Laric." Interview by Domenico Quaranta. *Art Pulse Magazine*, Web, July 23, 2011: www.artpulsemagazine.com/ the-real-thing-interview-with-oliver-laric/.

Reas, Casey. *Form + Design in Art and Architecture*. Princeton, NJ: Princeton Architectural Press, 2010.

Reifenscheid, Beate. "The Cosmos of Miao Xiaochun." *Miao Xiaochun: Macromania*, Ludwig Museum Catalog. Hong Kong: Timezone 8, 2010.

Rheingold, Howard. "Vernacular Video In Culture and Education." *Blip. tv*, October 9, 2008. Web, July 17, 2011: www.blip.tv/howardrheingold/ vernacular-video-in-culture-and-education-1348604.

Richard, Birgit. "Media Masters and Grassroot Art 2.0 on YouTube." In *Video Vortex Reader: Responses to YouTube*, ed. Geert Lovink and Sabine Niederer. Amsterdam: Institute of Network Cultures, 2008, pp. 141–52.

Richtin, Fred. *After Photography*. New York: W.W. Norton & Co., 2009.

Rock Hudson's Home Movies. Director Mark Rappaport. Waterbearer Films, 1992. DVD.

Roth, Evan. *Animated GIF Mashup Engine*. Created 2008; 2.0 Version Released 2010. Web, June 1, 2011: www.gifmashup.evan-roth.com/.

—"Ani Up: A Creation of the *Animated Gif Mashup* Engine." *Vimeo*, March 3, 2011. Web, June 1, 2011: www.vimeo.com/20903598 and www. networkcultures.org/wpmu/videovortex/archives/tag/evan-roth.

—"Ani Up: Animated Gif Mashup." *Vimeo*, March 10, 2011. Web: www.vimeo. com/20903598.

Ruangrupa. Web, September 25, 2011: www.ruangrupa.org/.

Rubal, Steve. "YouTube Uploads: 48 Hours of Video per Minute." YouTube, June 2011. Web, September 25, 2011.

Rush, Michael. *Video Art* (revised edn). London: Thames & Hudson, 2007.

Sack, Warren. "Picturing the Public." In *Structures of Participation in Digital Culture*, ed. Joe Karaganis. New York: Social Science Research Council, 2007, pp. 165–74. Web: www.ssrc.org/publications/ view/6A130B0A-234A-DE11-AFAC-001CC477EC70/.

Sahiner, Rafit. "An Attempt to Understand the Copy Artists' Works in Term of Ethics." Rıfat Şahiner Resmi Web Sayfası. Web, July 7, 2011: www. rifatsahiner.com/en/writings.html.

Said, Edward W. *Musical Elaborations*. New York: Columbia University Press, 1991. The Wellek Library Lectures at the University of California Irvine.

Sambrani, Chaitanya. "How and Away: Highways and Byways in Asian Art (2004)." In *Contemporary Art in Asia: A Critical Reader*, ed. Melissa Chiu and Benjamin Genocchio. Cambridge, MA, and London: MIT Press, 2011, pp. 163–78.

Sand, Olivia. "Shezad Dawood." *Asian Art News*, September 29, 2011. Web, September 29, 2011: www.asianartnewspaper.com/article/shezad-dawood.

Sasnal, Wilhelm. *Untitled (Elvis)* (2003). 16mm color film, 7 Minutes.

Sayre, John. *The Object of Performance: The American Avant-Garde since 1970*. Chicago, IL, and London: University of Chicago Press, 1989.

Schama, Simon. *Lanscape and Memory*. London: Harper Collins, 1995.

Schmidt, Kevin. *Epic Journey* (2010). HD video installation. Toronto, 2011: www. jmbgallery.ca/exhibitions.html.

Schneider, Nathan. "'This Is Just Practice' – The Story of the Wall Street Occupation." *YES! Magazine*, September 23, 2011. Web, September 25, 2011: www.yesmagazine.org/people-power/ this-is-just-practice-the-story-of-the-wall-street-occupation.

Schoenherr, Steven E. "Television Instant Replay." June 2, 2002. Web, May 2, 2006: www.history.acusd.edu/gen/recording/television8.html.

Scoggins, Lindsay. "Wonderland Mafia." YouTube, 2008. Web: www.youtube. com/watch?v=mkIoJZdKIAE.

Seale, Ansen. "Ansen Seale Photography – 'Panoramic Camera' Documentation." *Ansen Seale Photography – Documentation*. Web, February 2, 2011: www. ansenseale.com/statement.html.

Seaman, Bill. "Approaches to Interactive Text and Recombinant Poetics." *Electronic Book Review*, November 27, 2004. Web, July 21, 2011: www. electronicbookreview.com/thread/firstperson/languagevehicle.

"Search Engines Are About to Get a Whole Lot Smarter" (2002). Web, December 8, 2003: www.kartoo.net/e/en/artpresse/2002/wallstreet.html.

Second Front. "Grand Theft Avatar." Performance Documentation. Web: www. secondfront.org/Performances/Grand_Theft_Avatar.html.

Seid, Steve. "Chris Marker/MATRIX 168." In *Berkeley Art Museum and Pacific Film Archive*, 1995/6. Web, June 20, 2011: www.bampfa.berkeley.edu/ exhibition/168.

Sengers, Phoebe. "Practices for Machine Culture." Web: www.cs.cmu.edu/afs/ cs/user/phoebe/mosaic/work/papers/surfaces99/sengers.practices-machine-culture.html.

Sester, Marie. *Access. Documentation*. Ars Electronica, 2003. Web: www. accessproject.net/archive.html.

Shanken, Edward A. *Art and Electronic Media*. London: Phaidon, 2009. Themes and Movements.

—"The House That Jack Built: Jack Burnham's Concept of 'Software as a Metaphor for Art'." *Leonardo Electronic Almanac* 6.10 (November 1998). Web, July 31, 2011: www.mitpress.mit.edu/e-journals/LEA/ARTICLES/jack.html.

Sherman, Tom. "Vernacular Video." In *Video Vortex Reader: Responses to YouTube*, ed. Geert Lovink and Sabine Niederer. Amsterdam: Institute of Network Cultures, 2008, pp. 161–8.

Silent Movie (2005). Director Chris Marker. Video installation.

Silent Star Wars (2007). *Openculture.com*. Web, July 13, 2011: www. openculture.com/2011/07/star_wars_as_silent_film.html.

Simanowski, Roberto. "Digital Anthropophagy." *Leonardo* 43.2 (April 2010): 159–63.

—"Textual Objects: Refashioning Words as Image, Sound, and Action." Web: www.dichtung-digital.mewi.unibas.ch/.../Simanowski-Textual%20Objects. pdf.

Sittin' Pretty (1924). Director Leo McCarey.

Situationist International. "Détournement as Negation and Prelude," trans.

Ken Knabb. *Internationale Situationniste #3* (December 1959). *Situationist International Online*. Web, September 30, 2011: www.cddc.vt.edu/sionline/si/detournement.html.

Siufanga, David Tevita. "Untitled." *Vimeo*. Web, June 20, 2011: www.vimeo.com/14991008.

skynoise. **// *pixel rebooting shortly* //** (February 1, 2007): www.skynoise.net/2007/02/01/pixel-pirates-ii-soda-jerk-interview/.

SLC17. "Star Wars: The Empire Brokeback." YouTube, February 14, 2006. Web, September 10, 2011: www.youtube.com/watch?v=omB18oRsBYg.

Smith, Roberta. "As In Life, Timing is Everything in the Movies." *New York Times Online*, February 3, 2011. Web, June 6, 2011: www.nytimes.com/2011/02/04/arts/design/04marclay.html.

Soda_Jerk. *How I Learned to Stop Worrying and Love the Internet (HILTSWALTI)*. Web, September 29, 2011a: www.sodajerk.com.au/HILTSWALTI.php.

—*Pixel Pirate 2*. 2011b. Web, April 28, 2011: www.sodajerk.com.au/PixelPirate2.php.

Sollfrank, Cornelia. "Cornelia Sollfrank: Female Extension." *Media Art Net*. Web: www.medienkunstnetz.de/works/female-extension/.

Sontag, Susan. *Against Interpretation, and Other Essays*. New York: Picador, 2001 [1966].

—*On Photography*. New York: Picador, 1977.

SOUR, Masashi Kawamura, Qanta Shimizu, Saqoosha, and Hiroki Ono. *SOUR Mirror*. Computer software, *SOUR/MIRROR*. Web, July 31, 2011: www.sour-mirror.jp/.

Stallabrass, Julian. *Internet Art: The Online Clash of Culture and Commerce*. London: Tate, 2003.

Stallman, Richard. "The GNU Operating System." *Why Open Source Misses the Point of Free Software*. 2007. Web. 05 Mar. 2012. <http://www.gnu.org/philosophy/open-source-misses-the-point.html>.

Stangrunner. "U.S. Has the Lowest Piracy Rate in the World. Arrr..." Web log post, May 24, 2006. Web, June 2, 2008: www.iith096.blogspot.com/2006/05/us-has-lowest-piracy-rate-in-world.html.

Sterling, Bruce. "Closing Keynote: Vernacular Video," January 22, 2011. Web, July 17, 2011: www.boingboing.net/2011/01/22/bruce-sterling-talk.html.

"Steve Gibson. Interview." Online interview, July 14, 2011.

Stone, Allucquère Rosanne. "Sandy." In *The War of Desire and Technology at the Close of the Mechanical Age*. Cambridge and London: MIT Press, 1995.

Streitfeld, David. "Bargain Hunting for Books, and Feeling Sheepish About It." *New York Times Online*, December 27, 2008. Web: www.nytimes.com/2008/12/28/weekinreview/28streitfeld.html?_r=1.

Sundaram, Ravi. "Other Networks: Media Urbanism & the Culture of the Copy in South Asia." In *Structures of Participation in Digital Culture*, ed. Joe Karaganis. New York: Social Science Research Council, 2007. 49–72. Web: www.ssrc.org/publications/view/6A130B0A-234A-DE11-AFAC-001CC477EC70/.

A Supernatural Premiere (1995). Director Kevin Kelly.

Surowiecki, James. *The Wisdom of Crowds*. New York: Anchor, 2004.

Synthetic Reality. "Zhang Peili, *Last Words* (2003)." Web, September 21, 2011:

www.powerofculture.nl/uk/specials/synthetic_reality/exhibition_zhangpeili.
html.

"Tahrir Cinema." *Cinerevolution Now*, July 14, 2011. Web, October 5, 2011:
www.cinerevolutionnow.com/2011/07/tahrir-cinema.html.

Take The Square|Squares Around the World (2011). Web, October 5, 2011: www.
takethesquare.net/.

Tanner, Marcia. "Review: No Ghost Just a Shell." *Stretcher|Home*, 2002. Web,
October 5, 2011: www.stretcher.org/features/no_ghost_just_a_shell/.

Taring Padi. Web, September 25, 2011: www.taringpadi.com/.

Teh, David. "The Video Agenda in Southeast Asia, Or, 'Digital, So Not
Digital'." In *Video Vortex Reader II. Moving Images Beyond YouTube*,
ed. Geert Lovink and Rachel Summers Meyers. Amsterdam:
Institute of Network Cultures, 2011, pp. 162–77. Web, March
10, 2011: www.networkcultures.org/wpmu/weblog/2011/03/08/
video-vortex-reader-ii-moving-images-beyond-youtube-new-inc-publication/.

Templeton, Brad. "Hitler Tries a DMCA Takedown." *Brad Ideas Blog*, May 27,
2009. Web, July 20, 2011: www.ideas.4brad.com/hitler-tries-dmca-takedown.

Thacker, Eugene, "Protocol Is As Protocol Does." Foreword to *Protocol: How Control
Exists after Decentralization*. Cambridge, MA: MIT Press, 2004, pp. xi–xxii.

Thajib, Ferdiansyah et al. "A Chronicle of Video Activism and Online Distribution
in Post-New-Order Indonesia." In *Video Vortex Reader II: Moving Images
Beyond YouTube*, ed. Geert Lovink and Rachel Summers. Amsterdam:
Institute of Network Cultures, 2011, pp. 178–94.

Thalhofer, Florian, and Berke Bas. *Planet Galata*. Web, June 3, 2011: www.
planetgalata.com/.

Thalmair, Franz. "Amazon Noir Interview." *Cont3xt/furtherfield*.
Ubermorgan.com. *Amazon Noir: The Big Book of Crime*, 2008. Web,
July 14, 2011: www.perifericbiennial.wordpress.com/2008/09/18/
amazon-noir-the-big-book-crime/.

—"Amazon Noir Interview." *Furtherfield.org*. Web, August 10, 2011: www.
furtherfield.org/user/ubermorgen/amazon-noir.

Thomson & Craighead. *The Time Machine in Alphabetical Order* (2011). Web,
June 24, 2011: www.ucl.ac.uk/slade/slide/docs/thetimemachine.html.

Thornton, Mark. "Alcohol Prohibition Was a Failure." *CATO: Policy Analysis*
#157. July 17, 1991. Web, November 4, 3: www.cato.org/pub_display.
php?pub_id=1017.

Tofts, Darren, and Christian McCrea. "Many Happy Returns." *The Fibreculture
Journal 15: 2009 Remix* (2009). Web, April 28, 2011: www.fifteen.
fibreculturejournal.org/.

Tuters, Marc, and Karys Varnelis. "Beyond Locative Media." *Leonardo, Volume
39, Issue 4* (2006).

"Twitter." *Wikipedia*, January 3, 2009. Web, September 23, 2011: www.
en.wikipedia.org/wiki/Twitter#cite_note-51.

Ubermorgen, Paolo Cirio, and Alessandro Ludovico. "Amazon Noir: Big Book of
Crime. Documentation." In *AMAZON NOIR ~ The Big Book Crime*, November
2006. Web, August 12, 2011: www.amazon-noir.com/.

Vaidhyanathan, Siva. *The Anarchist in the Library*. New York: Basic Books, 2004.

Valdez, Sarah. "Open Source Art: Two Recent Exhibitions." *Art in America*, Bnet.

com, September 2005. Web, May 26, 2008: www.divus.cz/umelec/en/pages/ umelec.php?id=272&roc=2002&cis=4.

Van Buskirk, Eliot. "YouTube Search-and-Delete Code Makes Money for Rights-Holders." *Wired*, July 29, 2008. Web, September 19, 2011: www.wired.com/epicenter/2009/08/ how-copyright-holders-profit-from-infringement-on-youtube/.

Van Der Plas, Els. "Synthetic Realities." In *Synthetic Reality*. Hong Kong: Timezone 8, 2004, pp. 11–14.

Vernissage TV. "Rirkrit Tiravanija: Fear Eats the Soul at Gavin Brown's Enterprise, New York." March 18, 2011. Web, August 13, 2011: www.vernissage.tv/blog/2011/03/18/ rirkrit-tiravanija-fear-eats-the-soul-at-gavin-browns-enterprise-new-york/.

Virilio, Paul. *The Aesthetics of Disappearance*, trans. Philip Beitchman. New York: Semiotext(e), 1991.

—*The Vision Machine*. London: British Film Institute, 1994.

—"Critical Reflections." *Art Forum*, November 1995. Web: www.findarticles. com/p/articles/mi_m0268/is_n3_v34/ai_17865934.

—*Open Sky*, trans. Julie Rose. London: Verso, 1997.

—"Speed and Information: Cyberspace Alarm!" *CTheory.net*, August 27, 1995. Web, January 29, 2011: www.ctheory.net/articles.aspx?id=72.

Visa. "Travel Happy: Visa Commercial with Matt Harding." YouTube, December 3, 2008. Web, July 1, 2011: www.youtube.com/watch?v=TVOsBVDXSzc&feature=related.

Waltz, Mitzi. *Alternative and Activist Media*. Edinburgh: Edinburgh University Press, 2005.

Wang, Jianwei. "Is He a Traitor?" *Synthetic Reality*, 2002. Web, September 21, 2011: www.powerofculture.nl/uk/specials/synthetic_reality/exhibition_ wangjianwei.html.

—*Connection*. Two-channel digital video. Predates 2004.

Wang, Jin. *Brand New China: Advertising, Media and Commercial Culture*. Cambridge, MA: Harvard University Press, 2010.

Wang, Sue. "'Mona Lisa in the Weather from Sunny to Stormy' and 'Kitchen Servant Attacked by Sandstorms': Han Jingpen." In *CAFA ART INFO*. Beijing: Central Academy of Fine Art, October 4, 2011: www.en.cafa.com.cn/.

Wang, Tzzy. "China's Rocky Road to Combating Piracy Doesn't Dissuade it from Aspiring to Become an Innovative Society." *Ampersand*, August 6, 2007. Web, May 27, 2008: www.blogs.hillandknowlton.com/blogs/ampersand/ articles/9162.aspx.

Wardrip-Fruin, Noah. "Hypermedia, Eternal Life, and *The Impermanence Agent*." Web, February 17, 2005: www.impermanenceagent.com/agent/ essay.html.

Wardrip-Fruin, Noah, Adam Chapman, Brion Moss, and Duane Whitehurst. *The Impermanence Agent*. Web, September 29, 2011: www.impermanenceagent. com/agent/.

Wark, McKenzie. "Strategies for Tactical Media." *Tactical Media Files*. Web, June 14, 2011: www.tacticalmediafiles.net/article.jsp?objectnumber=46245.

Warren, Dan (as Mofolotopo). "Son Of Strelka, Son Of God (Narrated by Obama) – A Free Audio Story." *The Something Awful Forums*, May 10, 2004/

July 14, 2011. Web, September 10, 2011: www.forums.somethingawful.com/showthread.php?threadid=3422981.

Warren, Dan. "Son of Strelka, Son of God." In *The International House of Dancakes Blog and Podcast*, July 11, 2011: www.danwarren.blogspot.com/2011/07/son-of-strelka-son-of-god-audio-odyssey.html. Video, "Creation," animated by Ainsley Seago (June 25, 11): www.youtube.com/watch?v=Any38uNUelM; Video, "The Golden Age", animated by Adam Bozarth: www.youtube.com/watch?v=9PfrDjk0wP4&feature=related.

Wastring, Johan. "Visual Information Blog." In *Visual Information|Visualising Information*. Web, July 31, 2011: www.visualinformation.org/.

Watercutter, Angela. "*Life in a Day* Distills 4,500 Hours of Intimate Video Into Urgent Documentary." *Wired*, July 29, 2011. Web, September 3, 2011: www.wired.com/underwire/2011/07/life-in-a-day-interviews/.

We Tell Stories 6. Penguin Books, 2009. Web: www.wetellstories.co.uk/.

Wees, William C. "The Ambiguous Aura of Hollywood Stars in Avant-Garde Found-Footage Films." *Cinema Journal* 41.2 (Winter 2002): 3–18.

Weibel, Peter. Introduction. In *Drive*, by Jordan Crandall. Karlsruhe: Hatje Cantz, 2002, pp. 1–8.

—"The Post-Media Condition." In *Postmedia Condition Catalog*. Madrid: Centro Cultural Conde Duque, 2006.

Weibo, Sina. "500 Million Facebook Friends to Sina Weibo." YouTube, Web, July 11, 2011: www.youtube.com/watch?v=D9n24aP6lkA. (Viewable in China Only at Sina's site: www.bit.ly/qC6Cpa).

Weiser, Bill. "Ansen Seale's Photography Reveals Hidden Realities." *Lexjet Blog*, July 7, 2009. Web, February 3, 2011: www.blog.lexjet.com/?p=422.

Weiwei, Ai, and Lee Ambrozi. *Ai Weiwei's Blog: Writings, Interviews, and Digital Rants, 2006–09 (Writing Art)*. Cambridge, MA: MIT Press, 2011.

Whelan, Diane, and Jeremy Mendez. *This Land: Interactive* (2009). Web, September 5, 2011: www.thisland.nfb.ca/.

Whitaker, Iona. *Wang Jianwei: Through Multiple Lenses*, Catalog. Bejing: Ullens Centre for Contemporary Art, March 2011.

White Glove Tracking. Web: www.whiteglovetracking.com.

"Will YouTube Become a New Platform for Video Art? Guggenheim Experiments." *Art Radar Asia*, July 28, 2010. Web: www.artradarjournal.com/2010/07/28/will-youtube-become-a-new-platform-for-video-art-guggenheim-experiments/.

World Cup. "Where The Hell Is Matt? World Cup 2010". YouTube, May 4, 2010. Web, July 1, 2011: www.youtube.com/watch?v=TVOsBVDXSzc&feature=related.

—"Where the Hell Is Matt? World Cup 2010." YouTube, May 4, 2010. Web, July 1, 2011: www.youtube.com/watch?v=w9o5Pw1npAk.

Wu, Hao. "In Defense of Piracy." In *China and Globalisation. Marketplace: China*. American Public Media, ITunes Podcast, May 2, 2008. MP3.

Wu, Hung. "Miao Xiaochun and His *H2O*." *MiaoXiaochun.com*. Web, September 30, 2011: www.miaoxiaochun.com/yingwenyemian/ymxcdh2o.htm.

"Xeroxed Village: Chinese Secretly Copy Austrian UNESCO Town." *Spiegel Online*, June 16, 2011. Web, June 21, 2011: www.spiegel.de/international/europe/0,1518,druck-768754,00.html.

Xiaochung, Miao. *Microcosm* (2008). 14-minute animation.

Yates, Frances. *The Art of Memory*. London: Pimlico, 1992 [1966].

Yeo, Rob. "Cutting Through History: Found Footage in Avant-Garde Filmmaking." In *Cut: Film As Found Object in Contemporary Video*, ed. Stefano Basilico. Milwaukee, WI: Milwaukee Art Museum, 2004, pp. 13–27.

YouTube. "Content ID." YouTube. Web, September 19, 2011: www.youtube.com/t/contentid.

Zebonka. "Broke Trek." YouTube. May 27, 2010. Web, September 10, 2011: www.youtube.com/watch?v=7xSOuLky3n0.

Zhang, Ding. "Great Era." *ShanghART Gallery Shanghai*. On Felini. Digital video, 14 minutes, 2007. Web, September 25, 2011: www.shanghartgallery.com/galleryarchive/image.htm;jsessionid=A058F1E133D98A6CC1FE70AA0BF39056?id=10653.

Zhang, Do. *Great Era* (2007). Single-channel gallery installation, 14 minutes. Web, July 20, 2011: www.shanghartgallery.com/galleryarchive/image.htm;jsessionid=A058F1E133D98A6CC1FE70AA0BF39056?id=10653.

Zielinksi, Siegfried. "Discovering the New in the Old: The Early Modern Period as A Possible Window to the Future?" In *Miao Xiaochung 2009–1999*. Catalog. Cologne, Germany: Dumont, 2010.

Zizek, S. *Welcome to the Desert of the Real*. London: Verso, 2002.